Undergraduate Texts in Mathematics

Editors

S. Axler
F.W. Gehring
K.A. Ribet

Springer
New York
Berlin
Heidelberg
Barcelona
Hong Kong
London
Milan
Paris
Singapore
Tokyo

Undergraduate Texts in Mathematics

Anglin: Mathematics: A Concise History and Philosophy.
Readings in Mathematics.

Anglin/Lambek: The Heritage of Thales.
Readings in Mathematics.

Apostol: Introduction to Analytic Number Theory. Second edition.

Armstrong: Basic Topology.

Armstrong: Groups and Symmetry.

Axler: Linear Algebra Done Right. Second edition.

Beardon: Limits: A New Approach to Real Analysis.

Bak/Newman: Complex Analysis. Second edition.

Banchoff/Wermer: Linear Algebra Through Geometry. Second edition.

Berberian: A First Course in Real Analysis.

Bix: Conics and Cubics: A Concrete Introduction to Algebraic Curves.

Brémaud: An Introduction to Probabilistic Modeling.

Bressoud: Factorization and Primality Testing.

Bressoud: Second Year Calculus.
Readings in Mathematics.

Brickman: Mathematical Introduction to Linear Programming and Game Theory.

Browder: Mathematical Analysis: An Introduction.

Buskes/van Rooij: Topological Spaces: From Distance to Neighborhood.

Cederberg: A Course in Modern Geometries.

Childs: A Concrete Introduction to Higher Algebra. Second edition.

Chung: Elementary Probability Theory with Stochastic Processes. Third edition.

Cox/Little/O'Shea: Ideals, Varieties, and Algorithms. Second edition.

Croom: Basic Concepts of Algebraic Topology.

Curtis: Linear Algebra: An Introductory Approach. Fourth edition.

Devlin: The Joy of Sets: Fundamentals of Contemporary Set Theory. Second edition.

Dixmier: General Topology.

Driver: Why Math?

Ebbinghaus/Flum/Thomas: Mathematical Logic. Second edition.

Edgar: Measure, Topology, and Fractal Geometry.

Elaydi: An Introduction to Difference Equations. Second edition.

Exner: An Accompaniment to Higher Mathematics.

Exner: Inside Calculus.

Fine/Rosenberger: The Fundamental Theory of Algebra.

Fischer: Intermediate Real Analysis.

Flanigan/Kazdan: Calculus Two: Linear and Nonlinear Functions. Second edition.

Fleming: Functions of Several Variables. Second edition.

Foulds: Combinatorial Optimization for Undergraduates.

Foulds: Optimization Techniques: An Introduction.

Franklin: Methods of Mathematical Economics.

Frazier: An Introduction to Wavelets Through Linear Algebra.

Gordon: Discrete Probability.

Hairer/Wanner: Analysis by Its History.
Readings in Mathematics.

Halmos: Finite-Dimensional Vector Spaces. Second edition.

Halmos: Naive Set Theory.

Hämmerlin/Hoffmann: Numerical Mathematics.
Readings in Mathematics.

Hartshorne: Geometry: Euclid and Beyond.

Hijab: Introduction to Calculus and Classical Analysis.

Hilton/Holton/Pedersen: Mathematical Reflections: In a Room with Many Mirrors.

Iooss/Joseph: Elementary Stability and Bifurcation Theory. Second edition.

(continued after index)

George R. Exner

Inside Calculus

Springer

George R. Exner
Department of Mathematics
Bucknell University
Lewisburg, PA 17837
USA
exner@bucknell.edu

Mathematics Subject Classification (1991): 26-01, 26A06, 26A15

Library of Congress Cataloging-in-Publication Data
Exner, George R.
    Inside calculus / George R. Exner.
        p.   cm. — (Undergraduate texts in mathematics)
    Includes bibliographical references and index.
    ISBN 0-387-98932-3 (hc. : alk. paper)
    1. Calculus.   I. Title.   II. Series.
QA303.E95   1999
515 21—dc21   99-044785

QA
303
E95
2000

Printed on acid-free paper.

Production managed by A. Orrantia; manufacturing supervised by Jeffrey Taub.
Photocomposed pages prepared from the author's LaTeX files.
Printed and bound by R.R. Donnelley and Sons, Harrisonburg, VA.
Printed in the United States of America.

9 8 7 6 5 4 3 2 1

ISBN 0-387-98932-3 Springer-Verlag New York Berlin Heidelberg   SPIN 10744842

to my mother
Diana Campbell Exner

Introduction

Propaganda: For Students

There are plenty of things about calculus this text won't teach you, even though it *is* about calculus. Calculus is really two things: a tool to be used for solving problems for many other disciplines, and a field of study all its own. Calculus as a tool cares deeply about ways to find the largest value of a function, or obtain relationships between rates of change of some related variables, or obtain graphs of motion of physical objects. Recently introduced approaches for teaching calculus concentrate on the ability of calculus to provide models for many practical and physical situations (partly because the number of other disciplines that find calculus useful is growing all the time). All these are worthwhile and fantastically useful things to do. I regret to say that this text does none of this good stuff.

The study of calculus itself is really the internal, supporting structure, for all of the above tools and techniques. Why does a certain process produce the maximum value of a function on an interval, and is there a maximum value at all? Can we group functions into classes for which certain techniques work and others for which the techniques fail? What are the theorems we can state and prove about such classes of functions? These and similar questions are both hard and interesting; luckily, the calculator technology that makes learning calculus as a tool easier is equally helpful for tackling these matters.

To be honest, this material and these questions are hard enough so that very few people (nobody?) really understand them completely the first time around. Even for many people who go on to become mathematicians the understanding comes by bits and pieces as one works through first-term calculus, then second-term calculus, then third-term calculus, then a first real analysis course, then another analysis course, then the theory of integration, ...[1] At present, the teaching of calculus rather de-emphasizes these theoretical things, which means that when you (this means **you**) hit the first real analysis course you have to learn a great deal all at once. The purpose of this text is to teach you little bits of this theory along the way in order to make that course more comprehensible. You might not understand everything you read here fully, but the investment of your time is mostly for the future, so don't be concerned if everything isn't crystal clear — it would be amazing if it were. In this text, and in other courses, you'll revisit ideas over and over again, learning a little more each time.

Along with these matters, this text provides an introduction to the language in which mathematicians talk about theoretical matters. Theorems and proofs have their own language, and even before you study them formally it helps to have lots of examples. And even if the logical formalities aren't perfectly clear, you'll pick up a great many patterns that will be useful in future work. If things theoretical (proofs) aren't really what you like the best, be reassured that the approach here is a pretty gentle one. Again, we are laying a foundation for the future, not expecting perfect comprehension on the first presentation of this material.

One requirement for using this book is a willingness to put it down and work, even in the middle of sections. Perhaps your present pattern for reading a mathematics book is this: you race through each section spending some time (if any) looking at the examples worked out so as to be able to copy them later, and hurry to the assigned problems, which you do by imitating the worked examples and relying heavily on the answers in the back of the text (to the odd-numbered problems, of course; nothing can be done about the even-numbered ones!). Described this bluntly, this pattern sounds bad enough to be worth changing. The following icon indicates places where you need to put the book down and work some, and perhaps a lot, with pencil and paper, and often with a graphing calculator as well:

You will miss much of the value of this book if you don't do this every time. **Every** time. You are urged as well to work in groups, both in reading through the text and in working on the exercises, or at least to compare your work on the exercises with others. You may already be accustomed to (group) laboratory experiences in calculus, and such work is just as useful

---

[1] And after you teach real analysis you'll really have calculus down.

here. The point of both sorts of work is to get you away from watching while somebody else, usually the professor, does some mathematics. You are now at a stage where more independent work is necessary.

Propaganda: For Professors

The approach taken by this book is based on two beliefs. The first is that nobody (well, almost nobody) understands calculus fully (especially the theoretical parts) the first time around; multiple exposures are required. The second belief is that graphing calculators, now widely used to teach calculus in its aspect as tool, can be used as well to make the introduction of the theory of limits much easier for students. Anyone reading this book would by now have the right to expect it to be a calculus text, and, since it isn't, we turn next to the task of giving some background to explain what it is.

Recently calculus, or rather the teaching of calculus, went on the operating table for some major work. The "lean and lively" versions that emerged have altered the face of calculus teaching radically. Although the evidence is not all in, the author holds the opinion that some of the resulting approaches are a considerable improvement for many, even most, students. But more than fat was trimmed away; much of the underlying, supporting, theoretical structure was left on the table, too. Definitions of the limit and continuity, ϵ-δ work, and often statements of the classical theorems have disappeared in many of the new approaches, and even in more traditional approaches these topics have been withering away for years. For most students this is probably, or at least arguably, a good thing; for future majors in mathematics (and perhaps physics, engineering, and economics) I believe this is not at all a good thing.

Teaching or taking the first course in advanced calculus or real analysis in which students grapple with limits, and often at the same time as learning to prove, has never been exactly easy for anybody. Advanced calculus is a wonderful course in which far too many students previously successful in mathematics fail (or at least experience great frustration in achieving minimally acceptable results). The ideas are hard and proving things is also hard. A corollary of the first belief above is that investment in these theoretical ideas in the first three calculus courses can serve as a ramp up the cliff for students who will eventually take the more theoretical courses. This investment can come at many levels of sophistication: pictures of functions with a limit at a point and those without, pictures in which some particular values (perhaps numerical) of ϵ and δ appear, requiring students to find some numerical value of δ accompanying a numerical value of ϵ for some simple functions, carefully guided limit proofs for linear functions, ... A similar investment may be made in getting students used to the language of theorems, matters of hypothesis and conclusion, the construction of examples to illustrate the theorems and perhaps the necessity of the hypotheses, guided proofs of extremely simple results, etc. What level of investment is appropriate will vary from course to course based on student preparation, other course goals, and so on. But we often have two or three calculus courses lacking almost any investment of this kind. It seems on the borderline of sheer cruelty to thereby run students straight into the bottom of a cliff by expecting them to cope simultaneously with the notion of limit and the notion of proof, each for the first time, when we have boxed ourselves into needing comprehension to bloom immediately. Experiential bases for analysis concepts and mathematical language are both necessary and attainable.

We are lucky that an early investment in these concepts is made considerably less painful with the advent of graphing calculators. For a particularly potent example, consider the use of $\sin \frac{1}{x}$ as an example of a function with no limit at $x = 0$. I argue that before graphing calculators, this example was utterly unconvincing to most students; with a graphing calculator, students can zoom in to see the behavior about limits, which is impossible to pursue if drawing the graph of $\sin \frac{1}{x}$ is itself a real burden. For a more humble example, consider finding a value of δ to accompany the numerical value $\epsilon = .1$ for $\lim_{x \to 3} 4x$. The "trace" feature on any graphing calculator makes it easy to find a (convincing if unproved) value of δ. Further, it is no more difficult to attack a similar problem with $4x$ replaced by $e^{2x} \cos x^2$. Calculator technology makes the notion of "local linearity," so important to calculus as a tool, completely convincing; it can serve equally well to make finding δ strips to go with ϵ strips just as intuitive.

So what is this book? It is the theoretical pieces of introductory calculus presented, using appropriate technology, in a style suitable to *accompany* almost any first calculus text. It offers a large range of increasingly sophisticated examples and problems (almost always in numerical, graphical, and algebraic versions) to build understanding of the notion of limit and other theoretical concepts. It provides as well some introduction to the language of theorems, and practice in coming to understand what theorems say (and don't say) in ways not requiring proof (mostly by constructing examples and pictures). It is not aimed at all students, but rather those students (including mathematics students) who will study fields in which the understanding of calculus as a tool is not sufficient. Probably no teacher will choose to use all of the material here, but the text gives a range of potential investments for the future among which to pick and choose. The text uses, explicitly, the "spiral approach" of teaching, in which one returns over and over again to difficult topics, anticipating such returns across the calculus courses in preparation for the first analysis course; group work is useful and fits well with the design of the text. **Hints** are provided, but only for those exercises on which a student could go fundamentally, as opposed to locally, astray.

What isn't this book? This is not a calculus text: there is no discussion of finding maxima and minima, or "local linearity," or how to graph using the derivative, except inasmuch as these topics touch on theoretical matters. We don't discuss any of the computational procedures for derivatives except to prove that they (or at least some of them) work. In general, applications of calculus are omitted entirely, so this book is a supplement, not a replacement.

The text may also be used as the "content" text for a "transition to upper level mathematics" course. This text covers lightly the logical language of theorems, but probably not in enough depth for such a course; however, there are several texts presently in use (including the author's *An Accompaniment to Higher Mathematics*) that cover this area in depth but need some content area as practice field and to provide enough material for a full course. If the book is used in this way, a student should be ready to take quite a strong first analysis course, perhaps including metric spaces or other theoretical material difficult to include if one must start from the basics of limits.

Finally, the text may be used for students planning a career in teaching that will include some mathematics (or for other students not majoring in mathematics) as a very gentle advanced calculus course. Getting such students to appreciate the non-computational parts of mathematics can be a problem, and the informal style used here is of some benefit to those already nervous about mathematics in general and proof in particular.

A Note to All Parties

One goal of this text is to help students gain, along with other things to be gotten from calculus, an experiential base of concepts and formal language for the sake of future theoretical courses. This goal is, or at least was once, fairly traditional and customarily achieved with a much more economical presentation of material than that in this book. The spiral approach used here, in which the notion of limit (for example) is revisited repeatedly, the frequent use of "discovery" or "scratchwork" sections before proofs, and the use of icons calling for student work in the middle of the sections are nontraditional. In many ways the book is written as a dialogue between the author (really, the mathematics at hand) and the student reader. If the only goal were efficient presentation of material, this format would be both annoying (especially to a nonstudent reader) and inappropriate.

But the second goal is to begin to move students from being spectators at a presentation of mathematics to participating in the discovery and development of mathematics. Teachers often lament that students don't learn "actively," but a standard text is not a model for, nor does it serve to promote, active learning. The unusual characteristics of this text are there precisely to provide repeated opportunities for students to embark on this new way of learning mathematics, and to move, gradually, the responsibility for understanding new ideas to the students while giving them some tools to take on that challenge.

Therefore, successful use of this book requires some patience and willingness to try new things on everybody's part. The professor is encouraged not to race through the preliminary material on limits "since we don't really get down to business (real proofs) until later chapters." The student really does have to be willing to change, or at least try to change, many firmly embedded habits and beliefs about what it is to read a mathematics book. Comments to the author as to the success of the resulting experience are encouraged, from all involved.

Acknowledgments

My sincere thanks to my colleagues at Bucknell University, whose discussions with me about teaching have surely improved mine and helped shape parts of this text. I wish also to thank the editorial staff at Springer-Verlag, especially Ina Lindemann and reviewers of earlier drafts, who helped refine the text in many useful ways. Of course, the sins of omission and commission here are my own.

Finally, let me thank Claudia, Cameron, and Laurel, for everything.

Contents

1
Limits

A geometrical way to describe one of the fundamental concepts of calculus is the "slope of the tangent line" at some particular point on the graph of a function. Leaving aside for the moment such important questions as "is there a tangent line? ... only one?" and so on, one would like to compute the slope of that line in an efficient way. The first workers in (indeed, discoverers of) calculus thought of this slope in quite an intuitive way as the ratio of certain infinitesimal quantities (just as an ordinary slope is the ratio of some real numbers, namely the rise over the run). Putting this approach on a firm foundation turns out to be hard.

The standard way to make precise the intuition about the slope of a tangent line is to consider first some slopes of secant lines connecting two points on the graph of the function (one of them the point in question). None of these is the tangent line, but one examines the slopes of secant lines connecting the point to closer and closer points on the graph, hoping to detect a tendency of the resulting numbers. The goal is to extrapolate from the observed slopes of the secants what the slope of the tangent line "ought" to be.

The imprecision of this "search for tendencies" is made precise via the notion of "limit." We turn next to the basic definition, which we won't use in the context of slopes of tangent lines (the derivative) for quite some time.

1.1   The Definition of Limit

Here it is.

Definition 1.1.1 *Let f be a function defined on an open interval contain-ing the point a, except possibly at a itself. We say $\lim_{x \to a} f(x) = L$ ("the limit of f as x approaches a equals L") if, for every $\epsilon > 0$, there exists $\delta > 0$ such that for every x satisfying $0 < |x - a| < \delta$ we have $|f(x) - L| < \epsilon$.*

Yes. Well. Clearly something to be approached with caution. (For an even more intimidating symbolic form, see Definition 5.2.1 if you dare.)

This definition is, unfortunately, both important and difficult. It is cen-tral to analysis, let alone merely calculus. It is difficult in and of itself; it contains lots of "quantifiers" (whatever they are) and is at a high level of formality. Most people come to understand this definition and the concept it captures at various levels and through years of repeat encounters. The goal of a fair portion of this book is to give you a productive first encounter.

Modern approaches to teaching calculus often de-emphasize the formal side of this definition and concentrate on understanding its meaning. You've used techniques to try to understand this and other concepts used as tools for, say, modeling of physical situations (you may have been encouraged to think of things in three ways: numerical, graphical, and algebraic, for example). We'll use these techniques to grapple with the concept itself.

You may have worked with analogies or informal definitions: "the limit of f as x approaches a is L if when x gets close to a, $f(x)$ gets close to L." Something so vague couldn't possibly be wrong, but can't be a real definition either. We'll first try to push the informal sentence above toward more precision.

A crucial word in the above is "close." How do we decide when x is close to a or $f(x)$ is close to L? One problem is that close means different things to different people (say, a microbiologist and an astronomer). We need the mathematicians' version, which requires a detour into absolute value.

1.1.1   Absolute Value

The definition is the place to start.

Definition 1.1.2 *The absolute value of x, for x a real number, is written $|x|$ and defined by*

$$|x| = \begin{cases} x, & x \geq 0, \\ -x, & x < 0. \end{cases}$$

Convince yourself, by examples, that this thing is really doing what one tends to think of absolute value as accomplishing: returning the "size" of a number either positive or negative. Check also that the only number x such that $|x| = 0$ is $x = 0$. Further, and crucial for what follows, you need to convince yourself via numerical examples and pictures that the following proposition holds.

Proposition 1.1.3 *For any real number $r > 0$ and any real number x, one has $|x| < r$ exactly when $-r < x < r$. Therefore, the set of all x such that $|x| < r$ is exactly the set of all x such that $-r < x < r$, i.e., the open interval $(-r, r)$.*

In using absolute value to measure "closeness" (more generally, distance), we often face expressions like $|x - c|$, where c is some fixed number. The definition of limit has two such expressions, one measuring the closeness of x to a, and the other measuring the closeness of $f(x)$ to L. Some manipulations using the proposition above show that

$$|f(x) - L| < \epsilon$$

is equivalent to

$$-\epsilon < f(x) - L < \epsilon,$$

in turn equivalent to

$$L - \epsilon < f(x) < L + \epsilon.$$

You might prefer the interval notation for this, namely,

$$f(x) \in (L - \epsilon, L + \epsilon).$$

(Recall that "$\in$" means "is an element of.") Replace $f(x)$ by z for the moment, and observe that this is claiming that $|z - L| < \epsilon$ exactly when $L - \epsilon < z < L + \epsilon$. With a specific L and ϵ of your choice, verify this with some numerical examples of z. (Some pictures on the real number line might be good too.)

1.1:

We can work also with $0 < |x - a| < \delta$, if we recall $0 = |x - a|$ is the same as $x - a = 0$, i.e., $x = a$. So the condition

$$0 < |x - a| < \delta$$

is equivalent to the condition (actually a pair of conditions) "$x \neq a$ and $a - \delta < x < a + \delta$" or equivalently "$x \neq a$ and $x \in (a - \delta, a + \delta)$."

Using this equivalent notation, the definition of limit may be reformulated as follows.

Definition 1.1.4 *Let f be a function defined on an open interval containing the point a, except possibly at a itself. We say $\lim_{x \to a} f(x) = L$ if, for every $\epsilon > 0$, there exists $\delta > 0$ such that for every x satisfying $x \neq a$ and $x \in (a - \delta, a + \delta)$ we have $f(x) \in (L - \epsilon, L + \epsilon)$.*

In the graphical approach to the definition of limit (which we'll come to in a while), ϵ, δ, L, and so on appear naturally for the intervals in the alternate definition above. People sometimes use the language "an open interval centered at a" as a way to describe such intervals. The condition $x \neq a$ is crucial, but doesn't fit well into that language, so we may say "a <u>deleted</u> neighborhood centered at a" or "a <u>punctured</u> neighborhood centered at a" when appropriate.

1.1.2   Exercises

1.2: Sketch on the real number line the set of x's satisfying $|x - 5| < 1/2$. Sketch the related punctured neighborhood and express the set using absolute value inequalities.

1.3: Express the interval whose picture is below in both interval and absolute value inequality forms.

1.4: It will help for things we will do later to associate with an interval or punctured neighborhood of the real numbers a picture in the usual x-y coordinate system. The idea is simple if the interval is on the x axis: for each point x_0 on the x axis and in your interval, include in the diagram all points of the form (x_0, y) for any y whatsoever. Put differently, include the whole vertical line passing through each point of the interval (shading becomes useful!). If the interval in question is on the y axis instead, "stretch" it out parallel to the x axis instead. For each of the intervals or punctured neighborhoods in the exercises above, give the two-dimensional associated picture, first with the interval viewed as on the x axis and then with it viewed as on the y axis.

1.2   First Limit

We are now ready to tackle our first limit. We'll do only a little, for a simple function (f defined by $f(x) = x$), with ϵ set to .1 for example purposes (instead of "every $\epsilon > 0$"), and with "a" of the definition set to 3.

An improved informal definition will show what we need to do: informally, $lim_{x \to a} f(x) = L$ if we can make $f(x)$ as close to L as we want by making x close enough to a. With $\epsilon = .1$, "as close as we want to L" is "within .1". So we are trying to evaluate $lim_{x \to 3} x$, our measure of closeness is .1, and all we are missing is L. So the first question is, as x gets close to 3, what does x get close to (since, after all, $f(x) = x$)?

1.5:

This is indeed a "What color is Washington's white horse?" question, and so in this special case it is easy to identify L. But our special function is also a little confusing, because x is playing two roles, one as input to the function (standard) and one as output from the function (highly special). Each time you are faced with an "x" in this first example, take a second to figure out whether it is x as input or x as output.

OK, $L = 3$; we need to figure out how close to $a = 3$ we must make input x to get $f(x) = x$ within .1 of 3. In the formal definition, this means that, given our $\epsilon = .1$, we must specify our δ. Combining the formal and informal, we are being asked "how close (δ) must x get to 3 so that x is within .1 of 3?" (Which 3 is a? Which L? Which x is input, and which output?) All right, reread the question, and then answer it.

1.6:

Yes, of course, our special function makes life easy: if you are within .1 of 3, you are within .1 of 3. Formally, though, we have just chosen our δ (our measure of closeness of x to $a = 3$ in the domain of the function), which presumably guarantees that corresponding $f(x)$'s are close enough ($\epsilon = .1$ close) to our $L = 3$. All trivial, perhaps, but realize that you are now responsible for a great many x's in the domain of f. Indeed, this δ specified, you are responsible for all x's satisfying $0 < |x - 3| < .1$, meaning that you must make sure that each and every one of the $f(x)$'s for those x's satisfies $|f(x) - 3| < \epsilon = .1$. How many x's is that?

1.7:

Rather a big job. Try a few of the relevant x values numerically. Just for practice, graph on the real number line the set of x's your intuition made it seem simple to be responsible for.

1.8:

1.2.1   The Graphical Method

With the same example we illustrate a graphical approach to limits. Recall from Exercise 1.4 that we associated with intervals or punctured neighborhoods certain strips in the x-y plane. With each interval on the y axis, there

was associated a horizontal strip, and for each punctured neighborhood on
the x axis there was a (punctured strip? pair of strips? well, anyway, almost
a) vertical strip. On a piece of paper graph $f(x) = x$. Then, associated with
$L = 3$ and $\epsilon = .1$, draw the horizontal strip centered at $y = 3$ and whose
half-width is .1. Think of this as the "target" – any x whose $f(x)$ lands
in this horizontal strip is an x such that $f(x)$ is within $\epsilon = .1$ of $L = 3$.
Such an x you are willing to be responsible for, but there are lots of x's
to avoid ($x = 4$, for example) since their $f(x)$'s are not within $\epsilon = .1$ of 3.
The graphical way to say that is that, for example, $(4, f(4)) = (4, 4)$ is not
in the horizontal strip associated with $L = 3$ and $\epsilon = .1$.

Enter δ, the effort to limit your responsibility to x's whose $f(x)$'s are
properly placed. With $a = 3$ draw the set in the x-y plane associated with
the set of x such that $0 < |x - 3| < \delta$ (recall we try $\delta = .1$). Now one can
see quickly whether the δ chosen was sufficient to limit your responsibility
to x's (grouped around $a = 3$ and associated with the vertical strip) whose
$f(x)$'s are grouped around L, namely in the correct horizontal strip. In this
case, the simple $f(x) = x$, things seem to be fine.

Take a while to build, and study, the picture described above.

1.9:

1.2.2   Exercises

1.10: Continuing with the same function f, a, and L, redo things with
$\epsilon = .01$. Be explicit about your choice of δ, and draw the pictures.

1.11: Continuing with the usual function f, a, and so on, but with $\epsilon = .1$
again, suppose someone proposes $\delta = .01$. Is δ adequate? Draw the picture,
and decide using it and using some numerical x. What about $\delta = .05$ (still
with $\epsilon = .1$)? Write a brief paragraph summarizing your conclusions about
various possibilities for δ given a particular fixed ϵ.

1.12: Consider $\delta = .5$ for the usual function f, a, and so on, but with $\epsilon = .1$
again. Is δ adequate? Pictures and numerical examples should help decide.
If $\delta = .5$ is not adequate, there must be an x for which you are responsible
whose $f(x)$ is not where it is supposed to be. Give at least one explicitly.

1.13: Suppose that for the usual function but with $\epsilon = .1$ again, we return
to $\delta = .1$. We know that this choice of δ is adequate. Question: are you
responsible for $x = 2.9$? (Check the formal definition.) Would you be willing
to be responsible for the behavior of $f(2.9)$ anyway, or not? Repeat the
discussion for $x = 3$.

1.14: Change functions to f defined by $f(x) = 5$ (the constant function).
Take $a = 3$ again. What should $\lim_{x \to a} f(x)$ (i.e., L) be? Take again $\epsilon = .1$.

Can you find a suitable value for δ (graphically? numerically?)? Several suitable values for δ? An outrageous, but still suitable, value for δ?

1.3   Second Limit

Before moving on, let's see what we did, and mostly what we didn't do, with the simple case $f(x) = x$ above. We didn't cope at all with the "all $\epsilon > 0$" portion of the definition, but just with $\epsilon = .1$. We did produce a candidate for δ (in fact, several). Pictures and common sense showed that the δ we chose was adequate in that all the x's for which we were responsible behaved well, but we hardly proved anything. (We didn't examine $f(x)$ for each of the infinitely many x's, for example.) We'll repeat the process, fine for the time being as long as you aren't fooled into thinking that we have actually proved anything.

Consider the function f defined by $f(x) = 4x$ with $a = 3$. What is $\lim_{x \to a} f(x)$? (Note this is the same as "what's L?", because "L" and "$\lim_{x \to a} f(x)$" are both names for a number. It is true that the second of those names looks awfully complicated for the name of a single number, but that's what it is.) The number in question is luckily not too hard to guess.

1.15:

Let's make a point in passing. The good candidate for $\lim_{x \to 3} f(x)$ is not $f(3)$ merely by coincidence. But this is by no means a requirement of limits. Indeed, with a punctured neighborhood about a you are not responsible for a itself (that is, where $f(a)$ lands is not your problem). Functions whose limit at a and function value at a coincide are common, and frequently studied (put differently, we will tend to sweep other sorts of functions under the rug and try to ignore them). But function value as candidate for limit is common, not required.

The limit candidate identified, we ought to have to cope with all ϵ's, but again consider only $\epsilon = .1$ for the moment. Draw the picture in the x-y plane.

1.16:

We are missing δ. Our first example makes $\delta = \epsilon$ (both .1) a reasonable guess. Analyze it numerically and graphically, whatever your opinion is of this δ. Two things might happen. Perhaps this δ seems to work, in that all the x's in the punctured neighborhood of 3 are such that their $f(x)$ values lie in the right interval around L, or perhaps there are at least some x values which you would prefer not to be responsible for, because their $f(x)$ values are somewhere else. Which happens? Argue why $\delta = .1$ is satisfactory *or* give some bad x values allowed by that choice of δ.

1.17:

Well, yes. That choice of δ just doesn't seem to work.

Key Point Coming

It is vital to understand that you have only one trick in your bag (if $L = 12$ is right). You need another, better, candidate for δ. Argue why a δ larger than the one you already tried will never work; include a picture.

1.18:

Time to haul out the calculator and try some points to get a handle on an appropriate δ. Can you find the largest good δ?

1.19:

Check your value of δ via the graphical method; the function is so simple it may also be "obvious" what the largest possible value of δ is. Describe all values of δ you think likely to be satisfactory.

1.20:

Aside: Another Intuitive Definition

Another analogy people use for the limit at a point is a shooting/target one. You have a gun that cannot be adjusted perfectly, but can have its targeting parameters set to any tolerance you want. There is a target you must hit, not perfectly in the bulls-eye, but within some (small) region to be set by the judges. The process is then this: the judges set a region of closeness to the bulls-eye (analogous to setting ϵ). You must then set your targeting parameters to a tolerance (analogous to δ) adequate so that all your shots (analogous to x's in your δ-region) land in the region selected by the judges. If you can succeed in this task for any choice of region by the judges, you win (the function has a limit). This analogy works pretty well with the graphical method.

An alternate picture is sometimes useful. Instead of using the x-y coordinate plane, we'll use two copies of the real numbers, one for the domain of the function and one for the range. We get the following, with a, the proposed L, and ϵ filled in:

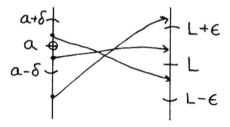

The advantage of this picture is that it is easy to draw, since you don't have to know anything about the real graph of the function, and it fits well with the "target" intuition of limit. The task, given ϵ, is as usual to find a value of δ so that all the arrows starting in the punctured neighborhood around a determined by δ land in the "target" region around L determined by ϵ. (Of course, we can't draw all the infinitely many arrows, one per input x, but as a representation it isn't bad.) Quick comprehension check: where is your lack of responsibility for what happens at a indicated in this model? We'll call this the "domain–range" picture in the future.
End Aside

1.3.1   Exercises

The following exercises are mostly designed to give you more examples of functions on which you can try numerical and graphical approaches to showing that limits are what you think they are, but also to disrupt intuitive conclusions based on too-simple examples. Intuition is great, but our past examples were so simple that almost any patterns you might think you have spotted are false. Here's a collection of more realistic examples. Your calculator is helpful, and remember its "TRACE" feature.

1.21: Consider the function f defined by

$$f(x) = \begin{cases} 4x, & x \neq 3, \\ -2, & x = 3. \end{cases}$$

What's $\lim_{x \to 3} f(x)$? (There might appear to be two choices – use first a numerical, and then a graphical, approach to choose between them.) With $\epsilon = .1$ what is an appropriate δ? With $\epsilon = .01$? If you are feeling bold and daring, conjecture a "formula" giving a successful δ in terms of the given ϵ. Show graphically and numerically that the wrong candidate for the limit fails. Try with the domain–range picture.

1.22: Consider the function f defined by $f(x) = -4x$. What's $\lim_{x \to 3} f(x)$? A δ for $\epsilon = .1$? With $\epsilon = .01$? Conjecture a "formula" giving a successful δ in terms of the given ϵ.

1.23: Continue with f defined by $f(x) = -4x$ and $a = 3$. In Exercise 1.22 you found δ for each of two ϵ's, $\epsilon = .1$ and $\epsilon = .01$. One value of δ is appropriate for each of these ϵ's. Which? Can you explain why, both in the language of responsibility for x values, and with a picture? In general, for some f, a, and ϵ_0, if δ_0 works, what works for some $\epsilon_1 > \epsilon_0$?

WARNING: In the previous examples there was misleading symmetry. Recall that given a δ you are responsible for all x in the open interval $(a, a+\delta)$ and also all x in the open interval $(a - \delta, a)$. (Split the punctured neighborhood into two pieces.) Let δ_r indicate how far from a it is safe to go to the right (values of x larger than a), and δ_ℓ the safe range to the left. They're the same for straight lines, but not in general. But the definition of limit forces you to choose δ so small that the symmetric above and below sets consist entirely of safe x's, even if this means discarding some additional x's that would be safe on one side or the other. Be warned.

1.24: Take f defined by $f(x) = x^2$ and $a = 2$. What's the limit? With $\epsilon = .1$, find δ. There are some pictures indicating the punishment for those who didn't read the warning above carefully and assumed symmetry; draw them, as well as a "successful" picture.

1.25: Continue with f defined by $f(x) = x^2$, but consider $a = 0$. What's $\lim_{x \to 0} x^2$? With $\epsilon = .1$, find δ. Consider also g defined by $g(x) = x$ near $a = 0$. Again, what is the limit? Further, say with $\epsilon = .1$, how does the largest possible δ for f (call it δ_f) compare with δ_g? Can you draw a picture making it clear why? Does this relationship between δ_f and δ_g (for the same ϵ) hold for other values of ϵ?

1.26: If you are willing to trust the technology on your calculator, finding δ to go with ϵ is not too difficult for any function. With f defined by $f(x) = e^x$, $a = 2$, and $\epsilon = .1$, find an appropriate δ.[1] What about $\epsilon = .01$?

1.27: (Technology again) Find $\lim_{x \to \pi/4} \sin x$, and, with $\epsilon = .1$, find a satisfactory value for δ. Use the sine function again at $a = 0$ and at $a = \pi/2$.

1.28: Consider the function f defined by

$$f(x) = \left(\frac{-x^3 + 6x}{5.656} \right)^{200},$$

and $a = 1.4$.[2] You may assume $L = f(1.4)$. Suppose we are working with $\epsilon = .01$. Work numerically: starting at 1.45 and decreasing by steps of .01, try to find a δ safe for values of $x > 1.4$ (that is, don't worry about values of x less than $a = 1.4$ for the moment).

---

[1]Note: for our examples with simple f and $\epsilon = .1$, $\delta = 1/1,000,000$ works, but you won't learn from it, so do honest labor.

[2]Yes, this is a cooked-up function to show something screwy. Play along.

You wouldn't have gotten this far in the educational system if you didn't realize that the value for δ indicated by finding a b such that $f(b)$ is close enough to L, and using that to give δ via $\delta = b - 1.4$, is not going to work. Now's the time for your graphing calculator: what happened?

Moral: when you pick a δ the responsibility for infinitely many values of x is substantial. With increasing or decreasing functions, finding one point that "works" seems to guarantee all those infinitely many nearer points (on that side of a) all at once. Not so with functions that are neither increasing nor decreasing throughout the interval.

1.29: The function in the previous exercise may have seemed too artificial. Here's another that shows a similar, but much worse, difficulty in coping with the "for all x" responsibility. Take f defined by $f(x) = x \cdot \sin(1/x)$ and $a = 0$ (note that we are really using the provision that the function need not be defined at the point). Take $\epsilon = .1$, and find a satisfactory δ. Hint: calculator; trace; graphical method. Transfer the picture to paper so you can fill in the details.

1.4   Anxiety, and Some Limits that Don't Work

The functions in the previous exercises show clearly why limits are hard. The function $x \cdot \sin(1/x)$ oscillates from positive to negative infinitely often near 0, yet still has a limit. That should make you nervous about how you would ever show that *all* the points in an interval were behaving. Also, you ought to distrust your calculator as a guarantor. After all, it graphs by plotting points and connecting the dots; lots of points, and accurately, yes, but only a finite number. (Think of the trace function – you really get only discrete choices of points as you run along the function, instead of a smooth range). What the function is doing between those points is really unknown. Perhaps your calculator missed some crucial jump or spike that happened over a very short interval, or maybe even lots of them. Calculators help investigate, not prove.

We turn here to functions that fail to have limits for less subtle reasons than the ones hinted at above. Proofs for *all* points in an interval will come later.

A starting point is what goes wrong when you pick the wrong candidate for the limit. For example, suppose you have f defined by $f(x) = 10x$ and $a = 4$. If you are momentarily befuddled, and decide L ought to be 30, you wind up in trouble. Show the picture for this with $\epsilon = .1$.

1.30:

Apparently, *none* of the values of x in any reasonable punctured δ-

neighborhood has $f(x)$ in the appropriate ϵ-interval around (the proposed) L. It's easy here to take the hint and fix the problem. Another wrong choice for the limit might occur with

$$f(x) = \begin{cases} x, & x \neq 4, \\ 0, & x = 4. \end{cases}$$

If you haven't made the break between limit and value of function, and try to use the value as the limit, you get in trouble again. Try it.

1.31:

Fair enough: you can't prove wrong is right.

In the examples above, there was a limit, and we picked the wrong one. If the idea of limit is interesting, there ought to be some functions with no limit. Find some from your library of functions.

1.32:

We'll start with one you might have found: f defined by $f(x) = 1/x$. Check that with $a = 2$, say, everything is calm.

1.33:

However, with $a = 0$ (the point you probably had in mind), things are quite different. Try $\lim_{x \to 0} f(x) = 1$ as a candidate; with $\epsilon = .1$, look for δ.

1.34:

By hand or by calculator, the initial δ you tried doesn't seem to work. That always means, of course, that there is some value of x in the punctured neighborhood determined by δ that has $f(x)$ outside the ϵ interval around L (in this case, 1). Give such an x. By itself that wouldn't mean anything; at least once in the past we've had to take two or more tries to obtain a working δ, even when there was one. But something's different here. What would improve if you took a smaller, even a dramatically smaller, choice for δ? In particular, for a δ that excludes the troublesome x from your first choice (good plan!), what happens?

1.35:

The point is that for the limit to fail, there has to be at least one value of ϵ for which we are unable to find a δ that works. To be really sure, we need more than a couple of failed δ's.[3] We somehow must be convinced that it is *impossible* to find a δ; no possible choice of $\delta > 0$ yields a punctured neighborhood of 0 for which all the $f(x)$ values are in the needed strip). Is this right?

1.36:

Here's an approach to something really close to a proof (by contradiction). If someone claims that $\delta = .1$ works, that person is responsible for $x = \frac{1}{2}$. Is this a safe x (i.e., is $f(x) = 1/x$ within .1 of 0)?

1.37:

All right, now somebody comes in and proposes that some particular δ_0 is satisfactory. All you know is $\delta_0 > 0$ and you are skeptical. From what you just did, guess a bad x for δ_0.

1.38:

To show that x is really bad, show that its $f(x)$ is not within $\epsilon = .1$ of $L = 1$, and, second, that it is one of the x's for which the person assumed responsibility with the proposed δ_0.

1.39:

These are almost true, and the form of the argument is exactly right. Ignoring a small subtle point for a moment, what has been done? We have shown that there is *no* successful choice of δ. So we showed that there was an ϵ (namely $\epsilon = .1$) such that no δ whatsoever was successful. This really constitutes a proof that the limit is not 1.[4]

The next question is, are we done analyzing $\lim_{x \to 0} 1/x$?

---

[3]Think back to $f(x) = x$, where we used $\epsilon = .1$ and found $\delta = .1$. The fact that a very confused person might have tried $\delta = 27$, $\delta = 519.3$, and $\delta = .5$, failing each time, doesn't mean that there isn't a δ. So a list of failures, even a long list, isn't the point.

[4]Well ...almost. For a particularly unfortunate choice of δ_0 (like $\delta_0 = 2$), $f(\delta_0/2)$ might land in the right strip. We'll cope with this later.

1.40:

Suppose someone claims instead that $\lim_{x \to 0} 1/x = 0$? Analyze this case to completion, destroying the foolish claim along the way.

1.41:

To show $1/x$ has no limit at all at 0, we have to convince ourselves somehow that there is *no* successful candidate for the limiting value. We won't pursue the details of this, but the pictures are pretty compelling. Here's a final question: consider the function f defined by

$$f(x) = \begin{cases} \frac{1}{x}, & x \neq 0, \\ 0, & x = 0. \end{cases}$$

Does this function have a limit at $a = 0$?

1.42:

1.4.1 Exercises

1.43: Consider the function f defined by

$$f(x) = \begin{cases} 1, & x > 0, \\ 0, & x = 0, \\ -1, & x < 0. \end{cases}$$

There are three obvious (not necessarily good) candidates for $\lim_{x \to 0} f(x)$. Argue completely and carefully for two out of three of these (your choice) that they can't be the limit. Hint: $\epsilon = .1$ will work just fine.

1.44: Continue with the function of the previous problem. Assuming that all three of the obvious candidates are ruled out , what remains to be done to show that the limit does not exist at 0? Do it for a typical example, as completely as possible.

1.45: Continue still with the function of Exercise 1.43. Observe that there *are* some values of ϵ (admittedly, large ones) with a working δ. Find such a value for ϵ, and a value for δ (several? outrageous?). Moral: the requirement that we can do the process for *all* positive values of ϵ is crucial; to be satisfied with "closeness" as measured only by some positive values of ϵ, would let some non-limits slip through.

1.46: (Continue the message of the previous exercise.) If we were to weaken the requirement that a δ can be found for all positive values of ϵ, and allow instead that it must be done down only to some small, but greater than zero, lower bound on the ϵ's, we get in trouble. Suppose the definition of limit said not "for any $\epsilon > 0$ ..." but "for any $\epsilon \geq .1$ ... ." Produce a function, by modifying the one in Exercise 1.43, with limit 0 at $x = 0$ by the "new" definition but not the old. What if the least ϵ were .01? If the least ϵ were ϵ_0? The point is that whatever the proposed least ϵ, there is a function with no real limit that slips through. The "for every $\epsilon > 0$" condition in the definition of limit is vital.

1.5   More Limits that Don't Exist

In this section we'll examine a striking example of a function without a limit, more than a century old but now less painful because of graphing calculators.

Define a function f by $f(x) = \sin(1/x)$ for $x \neq 0$, and any way you like at $x = 0$. Suppose it is claimed that $\lim_{x \to 0} f(x) = 0$. Analyze this proposal; what value for ϵ shows it won't fly? Can you do it numerically? Graphically?

1.47:

Things seem clear, calculator aided. Can we convince a calculator skeptic? Try this: The function seems to return to the value 1 (in fact, infinitely often, even in the interval $(0, 1)$, say). For what x does $\sin(1/x) = 1$?

1.48:

As before, a failed attempt at δ won't do; we must show that no possible δ works. This means showing that, for any δ, there is a point x in the punctured neighborhood around 0 determined by δ such that $\sin(1/x)$ is not within .1 of 0.[5] An x such that $\sin(1/x) = 1$ is a great candidate to show that some proposed δ was unsuccessful. Plot on the number line the various x's such that $\sin(1/x) = 1$; no matter what δ is suggested, is there one of these values in the interval $(0, \delta)$ (and so in the punctured δ-neighborhood around 0)?

---

[5]Of course, various δ proposals allow various x values to show δ fails. No single x must shoot down all δ's.

1.49:

Apparently, no matter what $\delta > 0$ is chosen, there is some one of the values $\frac{2}{(4n+1)\pi}$ in the $(0, \delta)$ interval. Supposing this to be true, draw the x-y plane picture to show that for any δ proposed, there are points on the graph of the function inside the vertical δ-strip but outside the horizontal ϵ-strip about 0 of width .1.

1.50:

This argument is essentially the proof that the limit is not 0. (How to write it formally comes later, but the ideas are all here.) For any value of δ, there is a promise made about the values of $\sin(1/x)$ for x in a certain punctured neighborhood of 0; we have produced a value of x for which the promise is broken. Thus no δ is satisfactory, and thus there is a value of ϵ with no satisfactory value of δ. Done.

Well, done in showing the limit isn't 0. Better candidates?

1.51:

Disposing of other candidates really divides into two cases. Suppose the candidate for the limit is anything but 1. For a well chosen ϵ, exactly the x's above shoot down any δ. Try it.

1.52:

If the proposed limit *is* 1, however, these values of x are not useful. After all, seems likely that if $\sin(1/x) = 1$, then $\sin(1/x)$ is in the horizontal ϵ-strip around 1 for *any* ϵ. What now? (A calculator graph may help.)

1.53:

Be explicit about ϵ and the bad values of x. Draw the picture to show that, because there is at least one bad x in any δ strip around 0, there is a value of $\sin(1/x)$ outside the horizontal ϵ-strip.

1.54:

It is useful to compare what we just did to show this limit did not equal zero with our work for f defined by $f(x) = 1/x$ (see Section 1.4). In each case we had to show that no δ would work for $\epsilon = .1$; in each case we did so by producing x in the punctured δ-neighborhood whose $f(x)$ was in the wrong place. Although the technical details of producing x varied in annoyance, the *form* of the argument is exactly the same. Check your work for Exercise 1.43 for the same form.

1.55:

Aside: Yet Another Intuitive Definition, and Localization

There is an alternative intuitive definition based on an industrial production model. You are in charge of a factory producing widgets of a certain length; you can't entirely control the length, but you can set your factory specifications to yield a length with as small an error as you desire. If so, and you are challenged by a customer to produce widgets of length L within an error of ϵ, you can set δ (your machine tolerances) so closely that all the widgets (x's) have lengths ($f(x)$'s) in the given interval. Note length exactly L (error, i.e., ϵ exactly 0) can't be done.

We stress also that the definition of the limit of a function at a point concerns the *local* behavior of the function. With δ picked (good or bad), you are discarding all values of the function for x outside the interval $(a - \delta, a + \delta)$. So what the function is doing "far away" from a is irrelevant to the limit. Indeed, if two functions coincide on an interval around a point a (even excluding a itself), then they will have the same limit at a (or both fail to have a limit at a).[6]
End Aside

1.5.1 Exercises

1.56: Define f by
$$f(x) = \begin{cases} 1, & x \text{ rational}, \\ 0, & x \text{ irrational}. \end{cases}$$

Start with the graph (your calculator refuses!).

There are two obvious (perhaps obviously wrong, but still the first guess) candidates for a limit at $a = 0$. Show that neither of them can be the limit by using the form of the argument above. What ϵ will provide a failure? For some specific numerical values of δ, produce an x for which you are responsible given your choice of δ, and for which $f(x)$ is outside the ϵ-strip.

Why can *no possible* value of δ work? Useful fact: any open interval contains both a rational number and an irrational number.

---

[6]Please remember this point, since it will be very useful later on.

1.57: Here's a variant of one of the functions above: define f by

$$f(x) = \begin{cases} \frac{1}{x} \cdot \sin(\frac{1}{x}), & x \neq 0, \\ 0, & x = 0. \end{cases}$$

Does this function have a limit at 0? If it does, argue graphically why, and try to find δ at least for various numerical ϵ's, and if possible for a general ϵ. If not, argue graphically and using the appropriate form of argument that it does not, and be as explicit as possible about δ's and values of x.

1.58: Repeat the previous exercise with another variant of one of the functions above: define f by

$$f(x) = \begin{cases} x^2 \cdot \sin(\frac{1}{x}), & x \neq 0, \\ 0, & x = 0. \end{cases}$$

1.59: Repeat with yet another variant of one of the functions above:

$$f(x) = \begin{cases} x, & x \text{ rational}, \\ 0, & x \text{ irrational}. \end{cases}$$

1.60: Repeat with f defined by

$$f(x) = \begin{cases} \frac{1}{x}, & x = 1/2^n, \text{ some } n = 1, 2, \ldots, \\ 0, & \text{otherwise}. \end{cases}$$

1.61: Continue with the function of Exercise 1.60. Does the function have a limit at $a = 1$? Show that it does graphically, and be specific about your value of δ for each value of ϵ. Does it have a limit at $a = 1/2$? Again be explicit about values of δ. Any points at which f has no limit?

1.62: Define a function "piecewise" as follows, where b is to be determined:

$$f(x) = \begin{cases} x, & x < 4, \\ -2x + b, & x \geq 4. \end{cases}$$

For what value(s) of b does this function have a limit at 4? What is that limit? With $\epsilon = .1$, what is a successful value of δ? With $\epsilon = .01$? Can you find a value of δ to go with a general ϵ?

2
Continuity

A crucial use of the idea of limit is to separate out a class of functions to study: exactly those whose value at a point coincides with the limit at that point. This is a useful class because it is relatively easy to study and it includes many familiar and useful functions.[1]

2.1 Continuity at a Point

The following definition should come as no surprise.

Definition 2.1.1 *A function f is <u>continuous</u> <u>at</u> <u>the</u> <u>point</u> b if it is defined in an open interval containing b and $\lim_{x \to b} f(x) = f(b)$.*

Observe that we have assumed f is defined on the sort of set that allows us to talk about the limit of f at b. Also, this is what it means for a function to be continuous *at a point*; the definition of f "continuous" (in some more broad sense) will be left until later.

It's time for some examples and non-examples. Almost always, the value part is easy and the limit part is the problem, so we steal from our work in Chapter 1.

We first studied f defined by $f(x) = x$ at $b = 3$. Well?

---

[1]Of course, they are familiar because we've swept the others under the rug. But more seriously, they were possible to study in the days before computers, and are used to model all sorts of physical phenomena.

2.1:

In an exercise, we tried f defined by $f(x) = 5$ (the constant function) with $b = 3$ again (this was 1.14).

2.2:

Next came f defined by $f(x) = 4x$ with $b = 3$.

2.3:

Then we tried a variant of the previous function:

$$f(x) = \begin{cases} 4x, & x \neq 3, \\ -2, & x = 3. \end{cases}$$

2.4:

You get the point: look at all of the examples in Chapter 1 and decide about continuity. The "weirder" ones (e.g., Exercise 1.59) can give surprising examples of functions continuous at a point. For some, remember that since "limit" is something which captures local behavior of a function, and the value at a point is surely local, a function might be pretty bad far away from a point and still have a limit there.

2.5:

(For some of the examples continuous at a point, the function is very bad far away from the point, somewhat bad a little closer, and increasingly "good" near the point.)

Even after you have reviewed all the previous examples, it seems prudent to analyze one function from scratch. Consider f given by $f(x) = x^2 + x$ at the point 2. Argue that the function is continuous at 2; so you need a limit, a function value, and equality of the two. Be explicit about the values of δ for $\epsilon = .1$ and $.01$. Graphically, can you find a value of δ for any ϵ?

2.6:

2.2   Naive Continuity vs. Continuity

We now contrast the definition of continuity with some non-definitions. One non-definition to set aside for the moment is the following: a function is continuous if you can draw it without picking your pencil up off the paper. Accept for the moment that this has to do with a function continuous on a set, and not just at a point. (Consider the function of Exercise 1.59, actually continuous at 0.) The non-definition above has to do with the function being continuous "everywhere," whatever that means. We'll pass on this for now.

Another "definition" frequently offered is that a function is continuous if it has no jumps, breaks, vertical asymptotes, or missing values. We can compare this to the real definition by considering the following question: suppose a function f has a jump at a point a; can it be continuous there? What if it has a break, asymptote, or missing value? While these questions aren't precise (really? what's a "jump"?), we can at least attack them graphically. Draw a picture of a function with your idea of jump; is it continuous there?

2.7:

"Break" is even less clear than jump; one possibility is what is shown below, where break is used in the sense of broken line (whatever that means). Is the function below continuous at the crucial point?

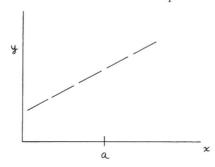

2.8:

Break might also mean jump, which we've already covered. A final possible meaning for break (or jump?) is illustrated in Exercise 1.21, where there is a value at the crucial point, but not where you would expect it. Can such a function be continuous at the special point?

2.9:

Note that this provides an example of a function not continuous at a point, but with a limit. With these friendly interpretations of break and jump, so far the intuitive definition is doing all right: if these things occur, continuity appears to be in trouble.

What about a function with a vertical asymptote?

2.10:

Here's a question: does providing the function with a value at the crucial point help things?

2.11:

Finally, consider a function with a "missing value" (excluding the vertical asymptote style), such as f defined by $f(x) = 4x$, $x \neq 3$. Observe that continuity fails right out of the gate, what with the "defined ..." condition and all.

2.12:

Conclusion: we should avoid the hazards in the intuitive definition.

Aside

We record some language: points of discontinuity where the limit is fine but there is a missing value or a value in the "wrong place" are sometimes called <u>removable</u> discontinuities. The name makes sense, since it is a trivial matter to change the function value at a single point and produce continuity. Consider our other examples of discontinuities in the non-definition we've been considering. Are they removable by defining or redefining the function at a single point?

2.13:

End Aside

With luck you are willing to agree that jumps, breaks, and so on are to be avoided. Unfortunately, most of the above is analysis of the wrong problem. We ought to have been analyzing the statement "If a function has no breaks, jumps, vertical asymptotes, or missing values, then it is continuous." In logical language (coming soon) this is of the form 'P implies Q' or 'if P then Q,' where 'P' is "the function has no breaks, jumps, vertical asymptotes, or missing values" and 'Q' is "the function is continuous." The claim is that any time 'P' is true, 'Q' must also be true. So we originally claimed that a function with no breaks, jumps, vertical asymptotes, or missing values is continuous. Unfortunately, what we actually did was show that if a function is continuous, it has no breaks, jumps, vertical asymptotes, or missing values. This is analysis of 'Q implies P.'

What's the difference: 'P implies Q,' 'Q implies P,' what's a difference in order between friends? But the difference is important. Suppose 'P' were "it is raining" and 'Q' were "I carry my umbrella." "If it is raining then I carry my umbrella" (an intelligent response to nature) is different from "if I carry my umbrella then it is raining" (a highly unusual power useful during droughts). Sometimes the truth of 'P' is enough to guarantee the truth of 'Q' ('P implies Q'), sometimes the truth of 'Q' is enough to guarantee the truth of 'P', sometimes each is enough for the other, and sometimes neither is enough for the other. Give some real-life examples of each of these situations.

2.14:

So back to the analysis we should have been doing, of "if a function has no breaks, jumps, vertical asymptotes, or missing values, then it is continuous." Is there some function out there which avoids all these problems, but still manages not to be continuous? Unfortunately, yes: define f by

$$f(x) = \begin{cases} \sin(\frac{1}{x}), & x \neq 0, \\ 0, & x = 0. \end{cases}$$

(Graph with your calculator and recall Section 1.5.)

2.15:

We've defined away the missing-value problem; there's no vertical asymptote; there's no "break" or "jump." Yet f is not continuous (at $a = 0$) since it has no limit there.

Remember what we are doing: we are trying to compare an intuitive definition of continuity with the real thing, and we've found a disagreement. While the intuitive definition may have been suitable once, it isn't any longer. But it did work to some extent: as a rule-of-thumb test for continuity of polynomials, ratios of polynomials (rational functions), and simple trigonometric, exponential, or logarithmic functions, it works fine. But to really understand continuity in general, and not some approximate version of continuity, the above isn't good enough. And the standard list of functions above doesn't include nearly all the functions out there.[2]

---

[2]Fact: most functions in the universe aren't even continuous. OK, but we

Here are a few exercises to work on your understanding of continuity at a point.

2.2.1 Exercises

2.16: Argue, using numerical values of ϵ (be explicit about δ) and by the graphical method, that f defined by $f(x) = xe^x$ is continuous at 2. Is there anything special about the point 2?

2.17: Consider f defined by $f(x) = \ln x^2$; based on a graph, at what points is this function continuous? Pick one, and argue for continuity graphically, and with two numerical values for ϵ. For all points of discontinuity, argue with some specific value of ϵ why the definition fails. Are there other reasons the function fails to be continuous there?

2.18: Define f by $f(x) = x^3 e^{\sin x^2}$. Consider $x = 3.5$; argue that f is continuous there. For $\epsilon = .1$, find an explicit value for δ. Realize that without a graphing calculator this would be a testing problem even for a professional mathematician. As long as you trust whatever numerical algorisms[3] are buried in the calculator, with a numerical value for ϵ this function is really no harder than x^2 and 3.5 is no harder than 0. You might think ahead, though, to coping with "for every $\epsilon > 0$," which the calculator doesn't help with at all.

2.19: Consider the function defined by

$$f(x) = \begin{cases} \frac{1}{x}, & x = 1/2^n, \text{ some } n = 1, 2, \dots, \\ 0, & \text{otherwise.} \end{cases}$$

(This function was considered in Exercise 1.60 and following.) At what points is this function continuous? Discontinuous? At what points of discontinuity does it have a removable discontinuity?

2.20: Define a function "piecewise" as follows, where b is to be determined:

$$f(x) = \begin{cases} x, & x < 4, \\ -2x + b, & x \geq 4. \end{cases}$$

For what value(s) of b is the function continuous at 4?

2.3 Continuity on a Set

There is a second (third?) use of the word continuity for functions: to capture the idea of a function which is continuous at a great many points

---

ought at least to consider *all* the functions we claim to be considering.

[3] "Algorithms" is more common. But see [2] for some of the (fascinating) history behind the mutation of the word over time.

(perhaps all points). The definition is unsurprising.

Definition 2.3.1 *Let f be a function and S a set. We say f is continuous on S if f is continuous at each point of S.*

This definition, together with the definition of continuity at a point, requires that f be defined at each point of S. In fact, more is required: f must be defined in an open interval about each point of S. Suppose, for example, that S were the closed interval $[0, 1]$; for f to be continuous on S, it would have to be defined in an open interval containing the point 1, and so the domain of definition would have to "spill over" outside of the set S. See Exercise 2.22 and following for more on this.

There is a symbolic form of this definition, which uses some symbols we will need soon anyway. We'd like to be able to talk about *all* points of a set S. The notation for "for all x" (or "for each x" or "for every x") is '$\forall x$'; to consider all points in S we might use '$\forall x \in S$'. With this notation, we can rewrite the definition above.

Definition 2.3.2 *Let f be a function and S a set. We say f is continuous on S if $\forall x \in S(f$ is continuous at $x)$.*

It is worth noting that, however written, this is a claim about all (usually infinitely many) x in the set S.

2.3.1 Exercises

2.21: For each of the following functions, decide whether it is continuous on the set $[-1, 1]$ and also whether it is continuous on the set $(0, \infty)$. If not, be explicit about where it fails to be continuous.

i) $f(x) = 4x$;

ii) $f(x) = 5$;

iii)
$$f(x) = \begin{cases} 4x, & x \neq 3, \\ -2, & x = 3; \end{cases}$$

iv)
$$f(x) = \begin{cases} \sin(\frac{1}{x}), & x \neq 0, \\ 0, & x = 0; \end{cases}$$

v) $f(x) = xe^x$;

vi) $f(x) = \ln x^2$;

vii)
$$f(x) = \begin{cases} \frac{1}{x}, & x = 1/2^n, \text{ some } n = 1, 2, \ldots, \\ 0, & \text{otherwise}; \end{cases}$$

viii)

$$f(x) = \begin{cases} x \cdot \sin(\frac{1}{x}), & x \neq 0, \\ 0, & x = 0; \end{cases}$$

ix)

$$f(x) = \begin{cases} \frac{1}{x} \cdot \sin(\frac{1}{x}), & x \neq 0, \\ 0, & x = 0; \end{cases}$$

x)

$$f(x) = \begin{cases} 1, & x \text{ rational}, \\ 0, & x \text{ irrational}; \end{cases}$$

xi)

$$f(x) = \begin{cases} x, & x \text{ rational}, \\ 0, & x \text{ irrational}. \end{cases}$$

2.22: (Continuity of functions on a closed interval) Consider some function like f defined by $f(x) = x$, $2 \leq x \leq 4$. It is not continuous at the points 2 and 4, since it is not defined in an open interval containing either of them. But it is doing as well as any function defined only on a closed interval could. So we make, in this and the following problems, some special definitions.

Definition 2.3.3 *Suppose f is a function and a is a point (not necessarily in the domain of f) such that f is defined on some open interval of which a is the right hand endpoint. We say that f has limit L from the left at a (or left-hand limit L at a), and write $\lim_{x \to a^-} f(x) = L$ if, for every $\epsilon > 0$, there exists $\delta > 0$ such that for all x in $(a - \delta, a)$ we have $|f(x) - L| < \epsilon$.*

Explore this definition. Pick some functions and try it out. What x's are you "responsible for" after you choose δ? What do the graphical and domain-range pictures look like? Suppose f is in fact defined in a punctured neighborhood of a and has a limit there — must it have a limit from the left there? Suppose a function is actually defined in a punctured neighborhood of a and has a left hand limit there — must it have a limit there? Give an example of a function and a point without a limit from the left.

2.23: (Exercise 2.22, continued) Formulate the definition of limit of a function at a point from the right (or right hand limit of a function at a point) by analogy with the definition in Exercise 2.22. The standard notation is $\lim_{x \to a^+} f(x) = L$. Explore with some more functions. Can you construct an example of a function with a right-hand limit of 5 at the point $x = 3$? Right-hand limit 5 and left-hand limit 2 at the point 3? Right-hand limit 5 and with no limit? A function with right-hand limit 5 and with no left-hand limit? A function with neither right-hand nor left-hand limit?

2.24: (Exercise 2.22, continued) Armed with the preceding definitions, we may give the following.

Definition 2.3.4 *Let f be a function defined on a closed interval $[a,b]$, with $a < b$.[4] We say that f is <u>continuous</u> <u>on</u> $[a,b]$ if f is continuous at each point of (a,b), $\lim_{x \to b^-} f(x) = f(b)$, and $\lim_{x \to a^+} f(x) = f(a)$.*

First, check that, for points in (a,b), we have the right situation for continuity at a point in terms of domain of definition of f.

Now try this definition out on some example functions defined on a closed interval. Your examples ought to include ones of the following types: nice quiet functions you expect to be continuous (are they?); functions you expect to be discontinuous because of points in (a,b) (are they?); functions with endpoint continuity problems, of all possible flavors.

2.3.2   Continuous Functions

There is another use of the word continuous, e.g., "f is continuous." What might this mean? There's one simple case: clearly, if f is defined on the whole real number line, then to say f is continuous ought to be (and is) to say that f is continuous on the set **R**. Can't be more continuous than that.

The (potential) difficulty comes when dealing with a function not defined on all of **R**. Suppose f is defined by $f(x) = 1/x$; faced with the graph,

2.25:

you might be surprised when I firmly announce that f is continuous. What could this possibly mean? There's a convention in force, which we record formally as a definition.

Definition 2.3.5 *We say a function f is <u>continuous</u> if it is continuous on the set consisting of its domain.*

Since the domain is a set (a subset of **R**), we have reduced the definition of "continuity" (just plain continuity) to "continuity on a set S" (for S the particular set consisting of the domain of the function).

Check that f defined by $f(x) = 1/x$ is continuous.

2.26:

Worry about the point $x = 0$ is fair, but since 0 is not in the domain of the function, it can't rule out continuity.

---

[4]This condition ensures that we have a "real" closed interval; if $a = b$, we have just a point, and most of the definition to follow makes no sense. It's unclear what it would mean for a function defined at a single point to be continuous (or discontinuous), nor does it seem worth pursuing.

Warnings

Often in elementary calculus, functions such as $1/x$ are called discontinuous. What happens is that the distinction between "continuous" and "continuous on a set" (and what set to use) isn't spelled out. This ambiguity *is* harmful and confusing. What's the precise thing to say? The function f defined by $f(x) = 1/x$ is continuous (i.e., continuous on its domain); it is not continuous on **R**; it is not continuous at $x = 0$.

There's another obligatory warning. The definition of continuous above, meaning continuous on the domain, clashes very badly with the intuitive definition of continuous we considered above having to do with "breaks, jumps, ..." Reason: often these "breaks, jumps, ..." occur at a point not in the domain of the function. As such, they destroy continuity at a point but not "continuity." A function can have lots of points of discontinuity and still be continuous *if each of them is at a point not in its domain.*[5]

Nail down these various uses with some examples of "continuous."
End Warnings

Here's one example to compare the definitions of continuous, continuous on a set, continuous on **R**, having points of discontinuity, and so on. Define f by

$$f(x) = \left\{ \begin{array}{ll} 1, & x > 0, \\ -1, & x < 0. \end{array} \right.$$

Just to be careful, what is the domain of f?

2.27:

Is f continuous at each point of its domain?

2.28:

Is f continuous? Is f continuous on the whole real line **R**? Does f have any points of discontinuity?

2.29:

Finally, can one define a function g such that $g(x) = f(x)$ for all points in the domain of f, but g is defined at $x = 0$ and is continuous there (recall that if so, the discontinuity is "removable;" see the Aside after 2.12)?

---

[5] Any phrase including "can have lots of points of discontinuity and still be continuous" is unpleasant; it's better if we rephrase it as "can have lots of points of discontinuity and still be continuous on a certain set." We simply don't punish a function on the basis of points at which it is not defined.

2.30:

Aside: A Moral of the Story

Perhaps it is this: the reputation of mathematicians for precision and lack of toleration of ambiguity isn't quite right. Nobody interested in complete lack of ambiguity would use the word continuous in so many different and potentially (even actually) conflicting ways.[6] But it is convenient to use a single word for all these various related situations, and we trust other mathematicians to sort out which definition of continuous is in use at the moment and detect if the ambiguity is troublesome.[7]

It might help to notice that you resolve ambiguities in mathematics all the time. How may solutions are there to $(x-1)^2 = 0$? Sometimes you say one, and sometimes you count the 1 twice. Which depends on context, and you choose the correct alternative almost without thinking. The ambiguities with continuity are others that you will learn to resolve effortlessly.

End Aside

The exercises following give you more chances to work with continuity of a function. After that, we would like to move on to theorems about continuous functions, but first will have to take a detour into a chapter about the language of theorems.

2.3.3   Exercises

2.31: For each of the following functions, decide whether it is continuous. If it is continuous, but has points of discontinuity (either in the domain or not), specify them carefully; are they removable?

i) $f(x) = 4x$;

ii) $f(x) = 5$;

---

[6]Fact: there are many more uses of "continuity" in more abstract settings, although they do give (usually) the kinds of continuity here when restricted to calculus situations.

[7]Some argue that this custom of allowing "harmless" or "resolvable" ambiguity makes the learning of mathematical language difficult. If a group of people who could write precisely agree among themselves to write imprecisely, students entering the field face things that, read literally, don't make sense. Complaints about the level of mathematical writing of students abound, but there is a case to be made that learning precise writing from imprecise models is pretty difficult. A discussion of the ambiguities and imprecisions in the use of "variables" we, professors and students alike, accept without thinking can be found in [3].

Unfortunately, none of this lets you off the hook. You still have to write precisely no matter what.

iii)

$$f(x) = \begin{cases} 4x, & x \neq 3, \\ -2, & x = 3; \end{cases}$$

iv) $f(x) = \sin(\frac{1}{x})$, $x \neq 0$;

v)

$$f(x) = \begin{cases} \sin(\frac{1}{x}), & x \neq 0, \\ 0, & x = 0; \end{cases}$$

vi) $f(x) = xe^x$;

vii) $f(x) = \ln x^2$;

viii)

$$f(x) = \begin{cases} \frac{1}{x}, & x = 1/2^n, \text{ some } n = 1, 2, \ldots, \\ 0, & \text{otherwise}; \end{cases}$$

ix)

$$f(x) = \begin{cases} x \cdot \sin(\frac{1}{x}), & x \neq 0, \\ 0, & x = 0; \end{cases}$$

x)

$$f(x) = \begin{cases} \frac{1}{x} \cdot \sin(\frac{1}{x}), & x \neq 0, \\ 0, & x = 0; \end{cases}$$

xi)

$$f(x) = \begin{cases} 1, & x \text{ rational}, \\ 0, & x \text{ irrational}; \end{cases}$$

xii)

$$f(x) = \begin{cases} x, & x \text{ rational}, \\ 0, & x \text{ irrational}; \end{cases}$$

xiii) $f(x) = \frac{1}{x}$, $x \neq 0$.

2.32: Consider f defined by $f(x) = 0$ for x in $[0, 1]$. Is f continuous?

2.33: Consider f defined by $f(x) = 0$ for x a rational number. Is f continuous?

2.34: Consider f defined by $f(x) = 0$ for x not an integer. Is f continuous?

3

The Language of Theorems

This chapter walks a thin line. We don't want to take the time to learn all of symbolic logic, as would be ideal to understand and prove theorems. But we have to get enough to work with, and without telling any lies or half-truths that might trip you up later. So the goal is modest: try to get enough of the language of theorems (actually the underlying language of mathematics) to understand some particular theorems of calculus.

Let's agree that a theorem is a guarantee of a certain kind. A theorem claims that if a certain thing or things happen (the hypothesis), then a certain thing or other things happen (the conclusion). Always. It's a guarantee special to mathematics, because it is not based on personal reputation ("George Washington never told a lie") or the laws of a state or country ("discrimination is illegal in the United States"). The guarantee's strength lies in the proof, which shows the conclusion always will hold if the hypotheses do, without exception. The statement "the exception proves the rule" indicates roughly that most real-life rules have exceptions, but to encounter one is so rare that the rule then comes to mind. In mathematics, the exception says that the proposed theorem is not a theorem.

What's the logical form of a theorem? If you know the answer perfectly, you probably don't need to be reading this, but try anyway.

3.1:

3.1 Implication

You probably said the form of a theorem is an implication (an "if-then"). That's a good place to start even though it is seldom exactly right. We must come to understand the pieces of an implication and what we are taking on when we claim (guarantee) in a theorem that an implication is true. We begin with the pieces, but recall that the logical notation for an implication is '$\Rightarrow$'.

Faced with an implication '$P \Rightarrow Q$' (or 'if P, then Q' or 'P implies Q'), recall (from geometry?) that there are names for the pieces P and Q: hypothesis and conclusion. Unfortunately, in a sense the names point out the roles of P and Q without saying what sort of things they actually are. What we need is the concept <u>statement</u>: an informal definition is "a sentence unambiguously either true or false."[1] Some examples of statements, in ordinary English and without looking for tricky meanings or stretching for ambiguities, are "Green is a color," "Green is a flavor," and "Air is 80 percent nitrogen." Note that these don't have to be true. Some "non-examples" are "She likes rain," "Don't do that," and "Hang in there." (The first is in some ways closer to being a statement than the others; hold that thought.) Construct some examples of statements and non-statements for yourself.

3.2:

Now it's easy: 'P' and 'Q' in a proposed implication '$P \Rightarrow Q$' must be statements. Even a natural English understanding of "if-then" indicates that "If hang in there, then she likes rain" has troubles. So an implication is built out of two statements and an 'implies.' The next question is, what sort of thing results? What kind of object is an implication '$P \Rightarrow Q$'? The key is that this is also a statement, a particular kind of statement built from two statements (presumably to capture some sort of relationship between them, but think formally for now). If they were connecting building blocks, you could take two statement blocks 'P' and 'Q' and connect them with an '$\Rightarrow$' block to get another statement.

If '$P \Rightarrow Q$' is a statement, there's a question. What?

3.3:

In ordinary language, we decide the truth or falsity of statements by thinking about what they mean. Logic, though, considers truth or falsity

---

[1]Note: this is informal, at least without firm definitions for "sentence," "true," and "false." We won't try to be more formal than this.

of statements without regard to meaning, but using only the true or false labels given to their component parts. The rules for labeling an implication true or false are completely formal and mechanical ones without regard to meaning or sense.[2] So think for a moment of simply giving labels 'T' and 'F' to statements, and how one ought to label the whole statement '$P \Rightarrow Q$' on the basis of the *labels* 'T' or 'F' given to its components (ignoring the meanings in favor of the labels). The device below, called a <u>truth table</u>, records the rule conveniently.

P	Q	$P \Rightarrow Q$
T	T	T
T	F	F
F	T	T
F	F	T

We'll start with the parts consistent with natural language examples. Consider "If it is raining, I will take you to the movies," and think of it in the sense of a guarantee or promise. Suppose first that "It is raining" has label 'T' (is true), and "I (will) take you to the movies" has label 'T' (is true — I did). I kept my promise, the guarantee was correct, and the implication should have label 'T.' Or suppose "It is raining" has label 'T' (is true), but "I (will) take you to the movies" has label 'F' (is false — I did *not*). You'd say I lied, that my guarantee was not upheld, and that the implication was false (had label 'F'), just as the table says. All is going well.

The last two rows are worse. In real life, if it is not raining ("It is raining" has label 'F' (is false)), we tend not to assign a truth or falsity to the implication at all (we somehow think of it as irrelevant or something). But to make '$P \Rightarrow Q$' a statement (true or false) we must make *some* assignment of label to the implication in the case in which the hypothesis has label 'F'. The agreement is to assign the implication 'T' in this case (that is, in both of the last two rows, whether the conclusion is assigned 'T' or 'F'). This is the benefit of the doubt. If it is not raining, you can't call me a liar, movies or no, so the guarantee holds.

Sometimes the result is strange. For example "If there are unicorns, then the moon is made of green cheese" has label 'T'. Produce (English language) examples of implications both true and false, covering all rows of the table.

3.4:

---

[2]Of course, the goal is to analyze real-world arguments, but that analysis is above and beyond the labeling rules.

3.1.1 *Exercises*

Assign truth values to the following sentences, or determine that they are not in fact statements.

3.5: If 24 is even, then 37 is odd.

3.6: If 24 is odd, then 37 is even.

3.7: If 24 is odd, then 37.

3.8: If 24 is odd, then 37 is odd.

3.9: If 24 is even, then 37 is even.

3.10: If 24 is odd, then n is even.

3.11: If the sine function is continuous at 2, then the sine function is differentiable at 2.

3.12: If the sine function is differentiable at 2, then the sine function is continuous at 2.

3.13: If a function f is differentiable at 2, then f is continuous at 2.

3.14: Continuity of the sine function at 2 implies differentiability of the sine function at 2.

3.15: Let P stand for the statement "42 is even," Q for the statement "31 is odd," and use the usual notation '$\neg P$' to stand for "42 is not even" and '$\neg Q$' for "31 is not odd." What are the labels, T or F, of the following?

i) $P \Rightarrow Q$;

ii) $Q \Rightarrow P$;

iii) $\neg P \Rightarrow \neg Q$;

iv) $(P \Rightarrow Q) \Rightarrow \neg Q$;

v) $(P \Rightarrow Q) \Rightarrow (P \Rightarrow Q)$;

vi) $(P \Rightarrow Q) \Rightarrow (Q \Rightarrow \neg P)$;

3.16: There are other ways to build new statements from old akin to those built using '$\Rightarrow$', each with its truth table. (Indeed, the "not" or "negation" appearing in Exercise 3.15 is one.) Here are the truth tables for "and" and "or;" explore them as you did "implies" above.

P	Q	P and Q
T	T	T
T	F	F
F	T	F
F	F	F

P	Q	P or Q
T	T	T
T	F	T
F	T	T
F	F	F

3.2   For Every ...

We turn now to why most theorems are not just implications. Theorems are also statements in the sense we just learned, necessarily either true or false. You might still think that the non-statement in Exercise 3.13, "if a function f is differentiable at 2, then f is continuous at 2" is a theorem and/or a statement. The difficulty is that the component "f is differentiable at 2" is not a statement (so the implication can't be one either). This differentiability claim surely depends on what f is, and you don't know. In this almost-statement, the symbol f acts as a pronoun, as "she" acted in "she likes rain" (see Section 3.1). If we fill in a value for that pronoun (e.g., Catherine the Great) it becomes a statement. Insert a particular function for f (e.g., the sine), get a statement. But with the pronouns in place, the components aren't statements (so the "theorems" aren't).

"But," you say, "no matter *what* function is inserted in for f, only one of two things can happen: either the function f_0 inserted for f is not differentiable at 2, so the hypothesis is false, so the implication is true by the peculiar truth table, or the function f_0 inserted for f *is* differentiable at 2, in which case (from calculus) f_0 is also continuous at 2. So the thing you are claiming isn't a statement really is, because no matter what is inserted for this pronoun, the resulting thing is true — can't get any truer than that."

This idea is almost right, but we need a way to make such an "almost-statement with pronoun" a real statement. The missing piece is in Section 2.3, where to say something about all points of a set we used '$\forall$' ("for all" "or for every"). Consider the following: "$\forall f$(if a function f is differentiable at 2, then f is continuous at 2)." Now this is a statement: either it really is true that for each function f, if f is differentiable at 2, then f is continuous at 2, or there is at least one f differentiable at 2 but not continuous at 2 (so the implication is false).[3]

A true statement with this 'for all' prefix is "$\forall x$(the sine function is continuous at x)." A false statement with such a prefix is "$\forall f$(if a function f is continuous at 2, then f is differentiable at 2)."[4] Construct more examples of true, and false, statements of this form (from your geometry past, perhaps?).

3.17:

We can now say why most theorems are not simply implications. Most geometry theorems, for example, have to do with classes of objects (e.g., right triangles) as opposed to single objects. Often the form is "for all

---

[3]Whether true or false it is a statement; as it happens, it's true.

[4]Translate the absolute value function over a bit for a counterexample.

objects of a certain type (if the object has properties A, B, ..., then the object has property C)." For example, "for all triangles (if the triangle has two congruent sides then the triangle has two congruent angles)." There are a single object theorems (for example, "$\sqrt{2}$ is not rational"), but comparatively few. Compare this to a related, and quantified, theorem: $\forall w(w$ is a prime integer $\Rightarrow \sqrt{w}$ is not rational).

This symbol "$\forall$" is called a <u>quantifier</u> (more precisely, the <u>universal quantifier</u>); statements involving a quantifier are said to be <u>quantified</u>. The use of quantified statements using "things that are close to statements but contain a pronoun (variable)" is part of the propositional calculus.[5] Many theorems will have the form "$\forall w(P(w) \Rightarrow Q(w))$," where $P(w)$ and $Q(w)$ are these almost-statements with pronoun w. For a while we'll try to avoid theorems with more than one pronoun and its quantifier. But the definition of limit exhibits a logical object (not a theorem, but a definition) with all sorts of quantifiers, including two uses of '$\forall$.' Also, theorems such as "If T and S are any two triangles ..." obviously call for two universal quantifiers, one for T, one for S.

Warning

The form of theorem above ("for all objects, property P guarantees property Q") is so common that often people are careless about making the quantification explicit. Here's a sample theorem.

Theorem 3.2.1 *If a function is differentiable at a point, then it is continuous at that point.*

Think a little. Is this a theorem with an explicit "for all" quantification?

3.18:

Easy enough. Here's the choice, then: either there is an implicit "for all" quantification, or "a function" was how the author chose to refer to a single object.[6] Common sense should eliminate one of these alternatives.

3.19:

It's unlikely that we've recorded a theorem about a single function, anonymous except to the author. You must, and will, get used to recognizing this abbreviated form of the universal quantifier. Words like "every,"

---

[5]There is another quantifier that we will get to soon.

[6]Parallel, perhaps, is "I wear a wedding ring." An arbitrary wedding ring? All wedding rings? No. There is a single, particular wedding ring which I've identified well enough in my own mind. But this won't do for theorems.

"each" and "any" are pretty clear cues to the universal quantifier; "a" used as above ("a function," "a triangle") is also often a cue. Another is a variable that appears out of nowhere: e.g., "If f is a function ..." is likely to mean "For any f, if f is a function ... ."
End Warning

3.2.1   *Exercises*

First determine whether the (proposed) theorem has a universal quantifier (implicit or explicit). If so, state clearly what sort of object is being universally quantified (that is, is the "pronoun" variable a function? an integer? ...). Finally, try to determine whether the proposed theorem is actually a theorem, meaning a *true* statement. You need not prove anything, but remember that the universal quantifier is a strong guarantee: it says that for *all* objects, something happens. Even a single object for which the claim fails dooms the proposed theorem as false and so not a theorem. So determining truth is hard, but determining falsehood is often easy.

3.20: For any real number x, if $x > 0$ then x has a real square root.

3.21: Any triangle with two congruent sides is equilateral.

3.22: Euler's constant is irrational.

3.23: Each positive integer is interesting.

3.24: If f is continuous on a closed interval, then f attains a maximum on that interval.

3.25: If f is a continuous function, and f is positive at some point and negative at another, then f is zero at some point.

3.26: If f is continuous on the closed interval $[a, b]$ and differentiable on the open interval (a, b), then there exists c in (a, b) such that

$$f'(c) = (f(b) - f(a))/(b - a).$$

3.27: The function f defined by $f(x) = e^x$ is continuous.

3.28: All polynomials are continuous.

3.29: If n is a positive integer, then f defined by $f(x) = \sqrt[n]{x}$ for $x \geq 0$ is continuous on $[0, \infty)$.

3.30: If g is any differentiable function and n is any positive integer, then

$$\frac{d}{dx}[g(x)]^n = n[g(x)]^{n-1} \cdot g'(x).$$

3.31: If g is a function defined by $g(x) = x^2 + 3x + 5$, then $g'(x) = 2x + 3$.

3.3   There Exists

We need the other quantifier "there exists," denoted '∃' (called the existential quantifier). We need it partly because it completes the list of quantifiers — '∀' and '∃' — but primarily we need it because it is crucial for definitions and theorems in calculus. The definition of limit requires that "there exists $\delta > 0 \ldots$." The Mean Value Theorem in Exercise 3.26 uses it obviously, and the Intermediate Value Theorem (Exercise 3.25) does even without the words "there exists" or obvious synonyms such as "there is." Yet more examples to come.

This quantifier is particularly useful for functions. A function f has associated with it a number of sensible concepts like the maximum point of a function (the x, if any, where the function achieves its largest value), an x where the function equals zero, or an x where the slope of the function is some particular value. It isn't clear how to say that a function has a maximum, but cleverness yields this: let $P(x)$ be the property that x is a maximum point for the function f. Perhaps there is no such point (define f by $f(x) = 1/x$, for example). But to say there is one is to say '$\exists x(P(x))$.' That is, "f has a maximum point" has been rephrased as "there exists a point which is a maximum point for f."

For another example, to say that an angle has an angle bisector is to say that *there exists* a half-line from the vertex of the angle inducing two congruent angles. To say that 2 has a square root is to say that there exists x such that $x^2 = 2$. Construct a few more.

3.32:

3.3.1   Exercises

By defining properties as needed (for example, property P above where $P(x)$ means x is a maximum point for f), write the quantified properties below in symbolic form. Two quantifiers may be needed.

3.33: ...there exists δ greater than zero ...

3.34: ...f has an x intercept ...

3.35: ...f has a maximum point ...

3.36: Every function f has an x intercept.

3.37: Every function f has a maximum point.

3.38: ...for every $\epsilon > 0$ there exists δ greater than zero ...

3.39: Every positive number has a square root.

4
Theorems about Continuous Functions

In this chapter we could ask either "Which functions are continuous?" or "What are the properties of continuous functions?" The first question rests on limits, so we will defer it until after Chapter 5 and concentrate on properties of functions continuous on a closed interval.[1] From past experience, or faith, or a peek at Chapter 7, we trust you are confident that there are such functions (e.g., polynomials).

Perhaps surprisingly, we won't prove these theorems. One sees the proofs of these theorems in a first course in real analysis or advanced calculus; there one grapples not only with limits at quite a technical level, but with crucial and somewhat subtle properties of the real number line. If you are like many students, you believe the real numbers consist of reasonable things like the rational numbers (fractions), $\sqrt{2}$ and the like, π, e, and ... and ... well, that's it. We won't attempt here the task of filling that out as is in fact necessary.

If not proofs, what's the agenda? First, we want to make sure that the logical structure of the theorems, quantifiers and all, is clear. Second, we want to construct enough examples fitting the theorems and non-examples not fitting the theorems so that the role of *continuity* in the theorems is clear (and other roles: see Section 4.3.1). This doesn't add up to an understanding of why the theorems are true, but it clarifies what the theorems "mean." In passing, we also point out some applications to show why anybody cares

---

[1] We will use here the special definition of continuity on a closed interval discussed in Exercises 2.22 and following.

about these theorems anyway.

Since we will be working with functions continuous on a closed interval, refresh yourself on Exercise 2.22 and following, especially Exercise 2.24.

4.1:

Finally, again just to remind you of something you already know, continuity of a function on a set is exactly (endpoint provisions aside) continuity of the function at a point (Definition 2.1.1) at all points of the set.

4.1   The Intermediate Value Theorem

We alter slightly the version from Exercise 3.25.

Theorem 4.1.1 (Intermediate Value Theorem) *Let f be a function continuous on a closed interval [a, b], and suppose $f(a) > 0$ and $f(b) < 0$. Then there exists x in (a, b) such that $f(x) = 0$.*

Start in on the form: What are the hypotheses? What is the conclusion? What quantifiers are present, and what are the variables they quantify? Is an implication present? What implication?

4.2:

Be clear that we assume three properties for the function, and that the "there exists" in the conclusion is not there by accident.

We next need to construct some simple examples of functions for which this theorem holds; clarifying the logical structure first lets us know what to do. We want a concrete example of a function f and a closed interval $[a, b]$ so that f is continuous on the whole interval and so that $f(a) > 0$ and $f(b) < 0$. Faced with such an example we can understand the conclusion of the theorem better: it claims that there exists an x with a certain property, and perhaps if we were staring at an example (picture?) we'd understand better what x has to do.

The natural inclination is to pick some a and b nice and simple (say, $a = 0$ and $b = 1$) and try to find f that does all the appropriate things. That's natural, but hard, since finding a function with three properties, including being positive at one prespecified place and negative at another prespecified place, is difficult. Try it, but stick to really simple functions.

4.3:

There's an easier way. The trick is to pick some function f and then try to locate some a and b such that a is to the left of b and $f(a) > 0$ and $f(b) < 0$. This is easier because fixing the function and then trying to locate an a and b is easier than fixing the a and b and trying to pick a function to fit. A little care choosing f is needed: what could go wrong with this program?

4.4:

Functions that avoid this difficulty include cubics; for specificity, let's suppose we've chosen f defined by $f(x) = x^3 - 4x + 2$. A graph of this function would be nice. Time to haul out your calculator.

4.5:

Based on the picture, there are, as it turns out, some delightfully simple (even familiar!) choices for a and b. What are they?

4.6:

This all in hand, what is the conclusion telling us? It claims that "there exists x in (a, b) such that $f(x) = 0$." If we specify to our particular situation (insert our particular f, a, and b) this means ...

4.7:

Is your intuition satisfied that x as claimed exists? Can you approximate its value? ("SOLVE" on your calculator might yield x such that $f(x) = 0$, but x is not in the interval (a, b). As far as the conclusion of the theorem goes, that isn't of interest. The theorem gives a point in the open interval, so you have to find that one (or ones? Hmmm ... a point to consider later). "TRACE" is a better approach.)

4.8:

Based on even this one example, the theorem should be clearer. To reformulate this theorem a little informally in graphical language helps too, yielding nice visual language at the cost of a little imprecision.

4.9:

The choice of function may have been surprising; what would we have done before the graphing calculator or without one? One choice is always an easier function, say, the sine function, which won't have any difficulty with being sometimes positive and sometimes negative. What are a good a and b, and resulting x?

4.10:

Repeat with $f(x) = x^3$, a more straightforward example of a cubic.

4.11:

That was indeed a curve ball; flexibility and common sense are required. When your example just won't fit the hypothesis, you have to either modify or discard. Modification is cheaper in terms of time and effort, so how about $f(x) = -x^3$? Similar modification would have been needed if your first try was x^2, say.[2]

4.12:

Our progress to this point, perhaps, is to understand some of what the theorem says, and in particular what the conclusion is insisting on.

Between Us

Here's a (helpful?) way to remember the point of this theorem. (You may leave me anonymous. No, really, I insist.) I think of it as the "Chicken Crossing the Road Theorem," for which it helps to view x as time t. If at one time a the chicken was on one side of the road, and at a later time b the chicken was on the other side of the road, there had to be a time in between when the chicken was actually crossing (actually on) the road. This assumes, of course, a continuously walking chicken, no trips circling the globe, etc., and doesn't attempt to say anything about why the trip was made. But it's easy to remember and captures the theorem quite precisely.

End Whispers

4.1.1   Exercise

4.13: Repeat the work above with f defined by $f(x) = e^x - xe^{-x} - 2$.

---

[2] Also, if trying x^3 hinted at of another version of the theorem, hold that (good) thought.

4.1.2 Why These Hypotheses?

We now have some understanding of what the theorem means. However, the examples above don't make it clear whether the continuity assumption is important. Could we guarantee the same conclusion without assuming continuity of the function on the closed interval (either assuming no continuity, or perhaps only continuity on the open interval (a, b))? To see if continuity is vital, we have to try using our examples of functions not continuous to see if we can evade the conclusion. For example, a function f defined on a closed interval $[a, b]$ but not continuous there, that *doesn't* satisfy "there exists x in (a, b) such that $f(x) = 0$" would show continuity is an important part of the theorem.

Review your examples of functions failing to be continuous at a point.

4.14:

Now here's the task: can you modify any of these functions (try graphically first) to produce f that fails to be continuous at even just one point of a closed interval, meets the rest of the hypothesis ($f(a) > 0$, $f(b) < 0$), and still evades the conclusion?

4.15:

Don't stop with one example. Is *any* of the various ways to fail to be continuous at a point enough to evade the conclusion of the theorem?

4.16:

One final thing: is discontinuity at a or b just as bad, so continuity on (a, b) still isn't enough?

4.17:

Since "continuity on a closed interval" is "continuity at a point" for *each* of the points of the interval (slight endpoint modification), your examples show that violation of this even at just *one* point of the interval is enough to evade the conclusion of the Intermediate Value Theorem. One bad point really causes a world of trouble; continuity is crucial.

Well, all right, continuity is needed, but what about the other two, $f(a) > 0$ and $f(b) < 0$? Is each of these needed — say, will assuming only continuity and $f(a) > 0$ but not $f(b) < 0$ allow evasion of the conclusion? Yes; construct some (easy) examples.

4.18:

Thus the Intermediate Value Theorem statement is efficient, with no hypotheses that may be done away with and still leave the conclusion guaranteed.

4.1.3 Generalizations

Perhaps you are impatient to point out that this statement of the IVT could be improved, as perhaps you noticed when trying x^3 as an example. True, x^3 is not positive for some point a and negative for some point b *with $a < b$*. Indeed, all the points where x^3 is positive lie to the right of all points where x^3 is negative. But pick a where x^3 is negative, and some point b (of course to the right of it) where x^3 is positive, and indeed x^3 is zero somewhere in between. Further, the picture shows it is for the same reason as in the stated theorem. Should the theorem be prejudiced in terms of positivity at a and negativity at b, as opposed to *vice versa*? Of course "$f(a)$ and $f(b)$ have different signs" is an improvement.

You might have done even better. Suppose f is continuous on $[a, b]$ and $f(a) = 1$ while $f(b) = -1$. Then f fits the original hypotheses, so there is x between a and b where $f(x) = 0$. Fine. Is there some point x_1 in between a and b such that $f(x_1) = 1/2$? Draw some sort of a generic picture.

4.19:

To require $f(x) = 0$ as opposed to $f(x) = 1/2$ seems hardly different, made clear by phrasing things in terms of points of intersection. To say $f(x) = 0$ is to say f intersects the line $y = 0$; to say $f(x) = 1/2$ is to say f intersects the line $y = 1/2$. Why single out the value 0? For *any* value in the interval $(f(b), f(a))$ ...

4.20:

Theorem 4.1.2 *Let f be a function continuous on the closed interval $[a, b]$ and such that $f(b) > f(a)$. Then for any z in $(f(a), f(b))$ there exists x in (a, b) such that $f(x) = z$.*

(You may fix this up to allow $f(b) < f(a)$ if you like.) Some books call *this* the Intermediate Value Theorem instead. From the original version one can prove all these apparently more general ones, so if you like the efficiency of the first formulation, remember it. If you prefer formulations without hidden generalizations, that's OK too.

4.1.4   Exercises

4.21: Use some formulation of the IVT to argue that there exists some number $\sqrt{2}$ by considering the function f defined by $f(x) = x^2$ and a useful interval of your choice. Picture? Other roots? Modify this approach to show every $d > 0$ has $x_1 < 0$ and $x_2 > 0$ so that $x_1^2 = x_2^2 = d$. Can you use this approach to argue that every positive number has only these two "square roots"?

4.22: Argue that each value c in the interval $[-1, 1]$ has some value x in the interval $[-\pi/2, \pi/2]$ satisfying $\sin x = c$. This allows one to define the "inverse sine" or "arcsine" function.

4.23: Show that the function f defined by $f(x) = x^3 - 2x + 1$ has a root in the interval $[0, .8]$.

4.24: A theorem is a guarantee in a positive sense, but not in a negative sense; if the hypotheses do not hold, the theorem *doesn't* say that the conclusion must also fail. Show this for the IVT: for example, given a and b and a function f that is *not* continuous on the interval $[a, b]$ and such that $f(a) > 0$ and $f(b) < 0$ can it still happen that there is an x in (a, b) such that $f(x) = 0$? Other hypothesis violations?

4.1.5   Who Cares?

Why bother learning this theorem anyway? It is of importance for future work (notably the Mean Value Theorem for Integrals), but perhaps "you'll know when you're older" is unsatisfying. Here is a practical application.

   Problem: given f continuous, find all x so $f(x) = 0$. Unless the function is unusually simple (a quadratic, say) solving exactly isn't an option. Your calculator and SOLVE is a good impulse. It's worth recognizing two things, though: first, that this is a recent option; many mathematicians from the 1700, 1800 and early 1900's would have given an arm for your $100 investment.[3] Second, *somebody* had to program your calculator to find that value. Actually, your calculator found an approximation to an x such that $f(x) = 0$. An understanding of the algorism is useful, and crucial for the "find *all* roots" task.

   A famous algorism for finding approximate zeros is known as the bisection algorism (or method). Find (somehow) points a and b with $f(a) > 0$ and $f(b) < 0$. Assume, to keep things simple, that a is to the left of b. By the Intermediate Value Theorem there is x between a and b such that $f(x) = 0$, so although you might be looking for a needle in a haystack at least there *is* a needle in the haystack. Now find the midpoint $m = (a+b)/2$

---

[3]Similarly recent is calculator graphing and tracing for a root.

of the interval $[a, b]$, and compute $f(m)$. If you are very$^\infty$ lucky you will find $f(m) = 0$, done. Otherwise, either $f(m) > 0$ or $f(m) < 0$.

If $f(m) > 0$, what about considering f on $[m, b]$ using the Intermediate Value Theorem?

4.25:

So a zero of f is trapped in a interval half the size of the starting $[a, b]$. Restart the process, finding another midpoint, and continue.

What about the case in which $f(m) < 0$?

4.26:

So again in this case the zero is trapped in an (albeit in a different) interval half the size of the one you started with.

Repeat the algorism, stopping when the interval is so small that any point is a good enough approximation to the x such that $f(x) = 0$. Graph some "generic" function with a zero and show a few steps of this algorism.

4.27:

This is a trivial computer or calculator program, and all you need is continuity of f and a way to evaluate f at a point. (Another method (Newton's method) usually converges faster but uses f', so f' must exist and be computable.) What can go wrong with the bisection algorism? By the Intermediate Value Theorem, nothing (leaving out operator error). At each step the function is positive at one end of the interval you are working on and negative at the other, so there is a zero in the middle.

What if there were *two* places where the function was zero in $[a, b]$? Draw the picture, and show some of the steps of the bisection algorism.

4.28:

What happens? At some point, you throw away an interval containing a zero (leaving yourself with another interval that also does). In a search for one zero, that's no problem. If you are trying to find all the zeros, understanding the algorism helps you see what you have to worry about.

4.1.6   Exercises

4.29: Use the bisection method to approximate to two decimal places a root of $e^{x^2} - 3e^x - 4$ in the interval $[0, 2]$.

4.30: Use the bisection method to approximate a value for $\arcsin 1/3$ (the discussion in Exercise 4.22 may be helpful).

4.2   The Maximum Theorem

4.2.1   Preliminaries

When you were solving max/min problems for calculus applications, you didn't worry that there might not be one. How optimistic! For the theorem, we need the definition of maximum point of a function on a set.

4.31:

Well, that's tougher than it looked: what should the maximum point of a function on a set mean? Perhaps a numerical example will help. Consider f defined by $f(x) = \log x \cdot e^{\sin x^2}$ on $[2,3]$. Ignore graphing by calculator temporarily and consider $x = 2.8$ as a candidate for a maximum point. Test by computing f at 2.8 and at four other points in the interval.

4.32:

There are two possibilities. Perhaps the function values at your points were less than or equal to $f(2.8)$. This doesn't guarantee 2.8 is a maximum point, but it is positive evidence. Perhaps, though, you tried a point x_1 such that $f(x_1) > f(2.8)$, showing immediately 2.8 is not a maximum point.

If f is the function and S the set, we want to say that a point x_0 in the set S yields the largest value of f on S; that is, if we consider *any* (hint, hint) other point x_1 in S, we have $f(x_1) \le f(x_0)$.[4] Try again for the definition.

4.33:

Observe, please, the universal quantifier. Define the maximum <u>value</u> of f on S to be the value of f at a maximum point.

Produce some examples of functions with, and without, maxima on various sets — first "with" on, say, $[0,1]$. Your graphing calculator should come in handy.

---

[4]Note: "maximum" or "largest" is meaning not "larger than any other" but "no other is larger." We would allow Mount Everest to be the "tallest" mountain even if it had a twin.

4.34:

What if the set is the whole real number line **R**: can you find some functions with maximum values? Without?

4.35:

It is harder to find functions without maxima if the set isn't **R**. Try $(0, 1)$ and some familiar functions (even just straight lines).

4.36:

The missing endpoint, which "wants" (?) to be the maximum point, prevents there from being one. Choosing points closer and closer to the endpoint yields larger and larger function values, but to suppose that one of them yields the largest is to be embarrassed by some point yet closer to the endpoint of the interval. This is one simple way that a function can fail to have a maximum on a given set.

That's one way to fail, but more dramatic is $\frac{1}{x}$ on $(0, 1)$.

4.37:

Yes, there is no maximum point in the sense that no point x_0 *in the set* has $f(x_0)$ largest, but in contrast to the last example, there is no largest value at all. These behaviors may be distinguished using "bounded above." A function f is <u>bounded above</u> on a set if there is some number M such that $M \geq f(x)$ for all x in the set (observe again the universal quantifier). Note that $M = f(z)$ for some z in the set is not required; that's the extra that makes for "has a maximum." Your examples show "bounded above but no maximum" and "not bounded above" are both possible.

4.2.2   Exercises

4.38: Find functions on the set **R**: a) with no maximum on **R**, b) bounded above but with no maximum, and c) not bounded above.

4.39: Define (carefully) "minimum point" and "f bounded below on S." Give examples as in the previous exercise on a variety of sets.

4.40: Find a single function f and three sets such that on one set f has a maximum point, on another set f has no maximum point but is bounded above, and on the third set f is not even bounded above.

4.2.3   Guarantees of Maxima

We began looking for maxima of f on a closed interval — no accident. Try to find a function with no maximum point on a closed interval.

4.41:

If you are unsuccessful, it is probably because you unconsciously stuck to continuous functions (discontinuous functions still feeling fairly new). But the following theorem shows this limitation doomed your efforts.

Theorem 4.2.1 (The Maximum Theorem) *Let f be a function continuous on a closed interval $[a, b]$. Then there exists a point x in $[a, b]$ that is a maximum point for f on $[a, b]$.*

(Recall that "continuous on a closed interval" has special endpoint provisions.)

Here's an important exercise. Look back at Section 4.1 for the process we used to understand the Intermediate Value Theorem. List the steps; what questions helped give us a handle on the theorem?

4.42:

The best possible thing to do is close the book and try each of those steps on your own in an effort to understand the meaning of the Maximum Theorem with the same process. The goal isn't a proof, but understanding the structure of the theorem and a rich collection of examples.

4.43:

Welcome back. We hope that you made it all the way through and just want to see what we did. (Better yet, compare examples with a friend.) Start with the hypothesis: we have a function continuous on a certain closed interval $[a, b]$. This means that f is continuous at a point in the usual sense at each (implicit universal quantifier in "each") of the points of (a, b) and is continuous in the special endpoint sense at a and b.

The conclusion is another existence claim: there exists (note the quantifier!) a point of $[a, b]$ that is a maximum point for f on $[a, b]$. (Yes, "maximum point" is itself quantified, but ignore the subdetails temporarily.)[5]

---

[5]Indeed, this is what definitions do. A complex condition gets a simple name, with details left for later as needed. Right now, "maximum point" is all the detail you need.

There are clearly parallels with the IVT, since in each theorem there *exists* a "special property" point in the domain.

In the IVT discussion we next collected some examples of functions that fit the hypotheses and tried to locate the special point. Example construction is easier for the Maximum Theorem; just pick some continuous function and an interval. Use your graphing calculator and various functions and closed intervals, and verify that there is indeed a maximum point.

4.44:

Just for another example, we'll consider $f(x) = \log x \cdot e^{\sin x^2}$ on $[2, 3]$ (note that only a very disturbed person would consider tackling this without a graphing calculator).[6] Graphing, zooming, and tracing as usual, ...

4.45:

my calculator seems to give a maximum point of about 2.81. The point is not to find this specific value but to show that, even for this ugly function, there is a maximum point in the interval. (Note that whatever the result of your four test points, 2.8 is *not* a maximum point. Moral: *all* other points in the relevant set must be considered.)

Collect two more examples: the first should show that two maximum points really can occur, so the scholarly discussion of "as large as" vs. "larger than" was appropriate.

4.46:

Second, construct an example to show that the closed interval condition is relevant. Can the maximum point actually occur at an endpoint?

4.47:

Next, do what we did in the "Why These Hypotheses?" section for the IVT: are the hypotheses are really needed to guarantee the conclusion? The process was to keep all but one of the hypotheses, violate the that one, and evade the conclusion. Here there is one obvious hypothesis. Can a function fail to be continuous on the closed interval and have no maximum point?

---

[6] Attempts via calculus techniques using the derivative fail on computational grounds.

4.48:

Will *any* type of failure of continuity (see Section 2.2) work?

4.49:

Can you place the point where continuity fails anywhere in the interval (in particular, at an endpoint) and still not have a maximum point?

4.50:

Conclusion: continuity of the function really does matter; without it, the guarantee of a maximum point can fail.

Another, less obvious, hypothesis was that the set is a closed interval. Is the choice of set crucial? Can you find an example of a function continuous on an open interval but without a maximum point?

4.51:

Here's a (slightly) harder problem. Intuitively, a "half-open" interval $\{x : a \leq x < b\} = [a, b)$ is "almost" a closed interval. Will the theorem necessarily hold with continuity and such a set?

4.52:

Here's one more. Perhaps the problem with $[a, b)$ is the obvious missing point; what about f continuous on $\{x : a \leq x < \infty\} = [a, \infty)$ (a half line)?

4.53:

So the choice of set appears crucial as well, since at least the obvious simple changes result in "theorems" that aren't actually theorems because they aren't true. As with the IVT, the Maximum Theorem is efficiently stated in that none of its hypotheses can readily be done away with.

For the IVT, we next tried to generalize its result.[7] How might we generalize the Maximum Theorem? Think hard about this one.

---

[7] Our generalizations could be proved from the original statement, so in a sense weren't more general, but displayed consequences hidden in the original.

4.54:

One thing always to note is an absence of symmetry; in the IVT, for example, one generalization came from realizing that the positive endpoint didn't have to be on the left. Here there is a different sort of asymmetry, namely maximum value of a function vs. a minimum value. Maximizing cost, for example, is rarely appropriate. State the minimum point version.

4.55:

(There's a combined max/min version too. Passing question: could a point be both a maximum point and a minimum point?) We will return later to proving the Minimum Theorem from the Maximum Theorem, but if one holds so should the other.

4.2.4 Exercises

4.56: For the Maximum Theorem, ask the question analogous to that in Exercise 4.24 about the IVT: if f has a maximum point on some set, is it required that f is continuous on the set and/or that the set is a closed interval? Dispose of this foolish hope completely in all its variants.

4.57: Here's a (useful?) combination of our two big theorems:

Theorem 4.2.2 *Let f be a function continuous on the closed interval $[a, b]$. Then f has a maximum point x_1 with corresponding maximum value $f(x_1) = \max(f, [a, b])$, and a minimum point x_2 with corresponding minimum value $f(x_2)$. Further, for each value z in $[f(x_2), f(x_1)]$, there is some x in $[a, b]$ such that $f(x) = z$. Thus the range of f on $[a, b]$ is a closed interval, in fact the interval $[\min(f, [a, b]), \max(f, [a, b])]$.*

"Proof": Apply the Maximum/Minimum Theorems to get x_1 and x_2. Then apply the Intermediate Value Theorem to the interval $[x_1, x_2]$ or $[x_2, x_1]$ (depending on which of x_1 and x_2 is the smaller; if $x_1 = x_2$, f is constant).[8]

Illustrate using $f(x) = \log x \cdot e^{\sin x^2}$ on $[2, 3]$.

4.2.5 Applications

You doubtless overlook one application of the Maximum Theorem. In calculus you find, via the derivative, maxima and minima of functions on closed intervals. It would be embarrassing, and a huge waste of time, to go

---

[8]Please remember this use of the IVT on a subinterval.

looking for the maximum of a function that had, in fact, no maximum at all. We'll eventually prove that if f has a derivative then it is continuous; the Maximum Theorem then tells you there *is* a maximum to be found.

There's another application important for theoretical work with limits. Surely a continuous function defined on a closed interval is bounded above and below, since there are numbers m and M (indeed, $\min(f, [a, b])$ and $\max(f, [a, b])$ from Exercise 4.57) such that each $f(x)$ is between m and M. When considering a product function $f \cdot g$, it often turns out that if f is "good" and g is bounded then $f \cdot g$ is "good" (see Section 5.2.2 for an example, including the meaning of "good"). And continuity of g on some $[a, b]$ is, by the theorem, enough to guarantee that g is bounded.

4.3   Digression: Compare and Contrast

The identical hypotheses of the Intermediate Value and Maximum Theorems (f continuous on a closed interval), and our examples and non-examples, suggest that the results depend in some way on the *same* combination of continuity and choice of set. A beautiful motivation for two of the important ideas of real analysis and topology is a collection of examples that shows this isn't right. We first push the Intermediate Value Theorem as suggested by Exercise 4.57.

Theorem 4.3.1 (Intermediate Value Theorem II) *Let f be a function continuous on some set S, and a and b points of S such that S contains the interval $[a, b]$. For any z between $f(a)$ and $f(b)$ there exists x in S (in fact, in the interval $[a, b]$) such that $f(x) = z$.*

"Proof": apply our original version of the IVT to the subinterval $[a, b]$.

But take a moment to check that if S is any interval-like object (an open interval, a half-open one, a half line with or without the endpoint included, or $\mathbf{R}$), it will contain the closed interval between any two of its points. Pictures are good enough.

4.58:

(Crucial) Moral: there is a sensible version of the Intermediate Value Theorem that can be stated for any interval-like set (e.g., those above), with f of course continuous. "Closed" or "finite-length" is not needed.

From our examples, however, the Maximum Theorem is *not* satisfied with open or infinite-length intervals. So it is counting on some property of closed intervals not shared by arbitrary intervals. For another example, consider S that is a union of two disjoint closed intervals, with f continuous on the whole set (i.e., on each closed interval). Can we guarantee that f has a maximum point on the set S?

4.59:

(Analogy: if we know that there *exists* a tallest tree in my neighborhood and a tallest tree in yours, then there must be a tallest tree in the neighborhoods combined, one of these two.)

For a numerical example, consider again $f(x) = \log x \cdot e^{\sin x^2}$, this time on the set which is the union of $[2, 3]$ and $[4, 5]$. What is the maximum point of the function on this set (do it graphically)? How does it compare to the maximum points of the function on the two sets separately?

4.60:

The reasoning isn't special to this function or these intervals.

Theorem 4.3.2 *Let f be a function that is continuous on a set S that is the union of two disjoint closed intervals. Then f has a maximum point in S.*

(Second, Crucial) Moral: Thus, while the Maximum Theorem obviously requires continuity of f on the set, the set can be some set (such as a union of two closed disjoint intervals) sufficiently "like" a closed interval.

Now let's try out the IVT out on the union of a pair of closed intervals. Construct an example to show that if you happen to pick point a in one of the intervals and point b in the other, there might be some m between $f(a)$ and $f(b)$ that is *not* (not, not, not) $f(x)$ for any x in the set S. (Note: an x <u>not</u> in S doesn't count!)

4.61:

4.3.1 The Payoff

This is it: the crux of the biscuit, the whole knockwurst, whatever. The Intermediate Value and Maximum Theorems are both naturally phrased as theorems about continuous functions on a closed interval. Both rely on continuity and on some property of the closed interval. However, since they may each be generalized to other sets, but not to the *same* other sets, they are relying on **different** properties of the closed interval. The development of language and concepts to distinguish these beautiful (and subtle) properties is the stuff of real analysis. Advertisement!

5
Limit Proofs

We return to the definition, recalled below, for a more thorough examination.

Definition 5.0.3 *Let f be a function defined on an open interval containing the point b, except possibly at b itself. We say $\lim_{x \to b} f(x) = L$ if, for every $\epsilon > 0$, there exists $\delta > 0$ such that for every x satisfying $0 < |x-b| < \delta$ we have $|f(x) - L| < \epsilon$.*

Look back to see what we did, and didn't do, with this definition.

5.1:

We stuck to simple functions. More importantly, often we found δ not for all $\epsilon > 0$ but for a few numerical values (e.g., $\epsilon = .1$). Occasionally we tried to give a "formula" yielding δ for any ϵ (such as $\delta = \sqrt{\epsilon}$); we showed some $\epsilon > 0$ had no good δ in Exercises 1.57 and following. But "for every $\epsilon > 0$ ..." was not really met. And when we found a (candidate for) δ, we avoided another infinite responsibility, to show that "for <u>every</u> x satisfying $0 < |x-a| < \delta$" something happened. Pictures and the calculator may have been convincing, but neither is a proof, and recall that calculators look at (many but) only finitely many points.

All this was an appropriate start, but we'll argue here that we can find δ for *every* ϵ, and that all x's such that $0 < |x - b| < \delta$ behave. We'll do this for appallingly simple functions (we'll work up to x^2) but we really will do it. But before that we must examine quantifier proofs at a very basic level.

5.1   Proof Templates

Good news: there is a proof template for the universal quantifier "for all" ('∀') and another for "there exists" ('∃'), each good enough for our needs.[1] "Template" here means a certain form (outline, pattern) to be followed with the details to be filled in; the empty spots are the same, the details problem-dependent.

5.1.1   Existence Template I

To prove something exists I must produce it and show it to you, and that is the basis of this template. To prove "there exists x with property P" ('∃$x(P(x))$)', I must

i) (∗) Find somehow my candidate for the good x;

ii) Commit myself publicly by announcing that this is my candidate;

iii) Prove to you that my candidate has property P.

The (∗) by *i)* is because although this step has to be performed, it is seldom in the proof. Off-camera scratchwork, or hope for inspiration, or a lucky guess produces the candidate. The proof contains an announcement of the candidate and the argument that it does indeed do what it should.

Suppose, for example, that it is to be shown that a particular line segment has a perpendicular bisector. In private we figure out how to construct the correct line. The proof, though, simply announces "the line constructed in such-and-such fashion is the perpendicular bisector." After this candidate announcement we prove (using congruent triangles or something?) that our candidate actually has the property of being the perpendicular bisector. Do that proof to see the template in action.

5.2:

To show "there exists $\delta > 0$ such that ..." requires this template. So there is always a bold, triumphant "Choose $\delta = .05$" or "Choose $\delta = \epsilon/2$" or "Let δ be the width of our family cookie jar" or whatever. After that comes the difficult part of proving that this δ works. (Announcing a candidate for δ is easy; the hard work is scratchwork to find a candidate for δ that will work.) How to find a good δ comes later. For now, understand that proving our δ works *must* be the goal.

---

[1] Indeed, they are adequate for almost all such proofs, here and elsewhere.

5.1.2 Exercises

In the exercises that follow, spot where the candidate is proposed, and locate the beginning and end of the "candidate is successful" proof. Pick out all of the pieces that are part of the existence proofs (ignore other things temporarily).

5.3: Let f be defined on $\mathbf{R}$ by $f(x) = 5$. Then $\lim_{x \to 2} f(x) = 5$.

Proof. We need to show that for every $\epsilon > 0$ there exists $\delta > 0$ with the usual property, so let $\epsilon > 0$ be arbitrary. Let $\delta = 1/2$. We must show that for every x so that $0 < |x - 2| < 1/2$ we have $|f(x) - 5| < \epsilon$. Let x be arbitrary so that $0 < |x - 2| < 1/2$; then $|f(x) - 5| = |5 - 5| = 0 < \epsilon$ using the definition of f and $\epsilon > 0$. So $\delta = 1/2$ works; since $\epsilon > 0$ was arbitrary, we are done.

5.4: Let f be the function defined on $\mathbf{R}$ by $f(x) = 4\sin x \cdot \cos x - 8\cos x \cdot \sin^3 x$. Then f has a maximum point on $\mathbf{R}$.

Proof. We claim $x = \pi/8$ is a maximum point. Observe that $f(x) = \sin 4x$ using some trigonometric identities. Then for any x_1, $f(x_1) \le 1 = f(\pi/8)$. Since x_1 was arbitrary, $\pi/8$ is indeed a maximum point.

5.5: Let f be defined on $\mathbf{R}$ by $f(x) = x^2$. Then $\lim_{x \to 3} f(x) = 9$.

Proof. We show that for every $\epsilon > 0$ there exists a $\delta > 0$ so that if $0 < |x - 3| < \delta$ we have $|f(x) - 9| < \epsilon$, so let $\epsilon > 0$ be arbitrary. Let $\delta = \min(1, \epsilon/7)$. We must show that for every x such that $0 < |x - 3| < \delta$ we have $|f(x) - 9| < \epsilon$. Let x be arbitrary such that $0 < |x - 3| < \delta$; then $|f(x) - 9| = |x^2 - 9| = |(x - 3)(x + 3)| = |x - 3| \cdot |x + 3| < |x - 3| \cdot 7 < \delta \cdot 7 \le \epsilon/7 \cdot 7 = \epsilon$, where the first inequality uses $|x - 3| < \delta \le 1$ and hence $x < 4$. So δ is as required; done, since $\epsilon > 0$ was arbitrary.

5.6: For any function f, there exists a function g such that $f + g = 0$.

(Given two functions f and g, $f + g$ is defined by $(f + g)(x) = f(x) + g(x)$ for all x. Also, h and i are equal if $h(x) = i(x)$ for all x. Finally, "0" here denotes the zero function z: $z(x) = 0$ for all x.)

Proof. Let f be arbitrary. Define g by $g(x) = -(f(x))$ for all x. We need $f + g = 0$, i.e., $(f + g)(x) = 0$ for all x. Let x be arbitrary; then $(f + g)(x) = f(x) + g(x) = f(x) + (-f(x)) = f(x) + -f(x) = 0$.

5.1.3 Existence Template II

Actually the template presented next is not really different from the first one for existence proofs. To prove '$\exists x(P(x))$' is still to

i) (∗) Find somehow a candidate for the good x;

ii) Commit ourselves publicly by announcing that this is the candidate;

iii) Prove that this candidate has property P.

Finding the candidate is the change to the template; our constructions and computations are replaced by another candidate source.

Consider what happens if we have an existence statement as part of the *hypothesis* (to use, not to prove). How can we use it? That is, *given* '$\exists y(Q(y))$' and to prove '$\exists x(P(x))$', how do we get something useful out of '$\exists y(Q(y))$'?

Example: prove every acute angle θ has an angle bisector, using the fact that every acute angle has an inscribed circle. To display all the existence statements we write this: "Given: $\exists C(C$ is a circle inscribed in $\theta)$. To prove: $\exists \ell(\ell$ is an angle bisector for $\theta)$." Here's scratchwork to find a good candidate for the angle bisector (you supply the diagrams).

Scratchwork   Let A be the vertex of angle θ. Introduce the circle C inscribed in the angle (which exists by the hypothesis). Let B be the center and D and E the points of tangency with the half-lines forming angle θ. The line through A and B is (?) the angle bisector. To prove that, we'll probably use congruent (similar?) triangles, so get triangles by drawing radii BD and BE. Are the two obvious triangles congruent? Their common segment is surely congruent to itself, and the two radii are surely congruent. What about the third sides? Not obvious. Angles? Lines tangent to a circle are perpendicular to the radius, so $\angle ACB$ and $\angle ADB$ are congruent. But "angle-side-side" isn't any good. Rats. Oh, but angles $\angle ACB$ and $\angle ADB$ are more than congruent, they are *right* angles, so the triangles are really right triangles. Pythagorean theorem! So DAB and EAB are congruent.

Proof   Let A be the vertex of angle θ. Introduce the circle C inscribed in the angle that exists by the hypothesis, and let B be its center. We claim that ℓ, the line through A and B, is an angle bisector for θ. To show this, let D and E be the points of tangency of C with the half-lines forming θ. Angles $\angle ACB$ and $\angle ADB$ are right angles, and so triangles ACB and ADB are each right triangles with hypotenuse AB. Since AB is congruent to itself, and DB is congruent to EB since each is a radius, by the Pythagorean theorem we have AD congruent to AE. Thus triangles ACB and ADB are congruent, so corresponding angles EAB and DAB are congruent. So ℓ is the angle bisector as desired.[2]

Our candidate for the angle bisector was not built in some specific or numerical sense (like $\delta = .05$); we took something we knew existed and used it to produce our candidate. The output of the existence hypothesis was an inscribed circle and it, or something gotten from it (a particular line through its center) was our candidate. This is absolutely typical of how existence hypotheses get used.

We can use this strategy with some old friends from Section 4.2.3.

---

[2]Note the difference in tone and content between scratchwork and proof.

Theorem 5.1.1 (The Maximum Theorem) *Let f be a function continuous on a closed interval $[a, b]$. Then there exists a point x in $[a, b]$ that is a maximum point for f on $[a, b]$.*

Recall also the Minimum Theorem:

Theorem 5.1.2 (The Minimum Theorem) *Let f be a function continuous on a closed interval $[a, b]$. Then there exists a point x in $[a, b]$ that is a minimum point for f on $[a, b]$.*

Here's scratchwork for a proof of the second from the first.

Suppose f is continuous on $[a, b]$ and we want a minimum point. Consider g on $[a, b]$ defined by $g(x) = -f(x)$ for all x in $[a, b]$. Observe that g is continuous on $[a, b]$.[3] But then the Maximum Theorem guarantees us ...

5.7:

The Maximum Theorem gave us, say, x_* (remember, as applied to g, not f). We want a candidate for a minimum point for f, however; does x_* look good? Consult your graphs.

5.8:

Fantastic! We have a candidate ... .[4]

Again the output of one existence result gave the candidate for an existence proof. (This time, the object itself, not an offshoot. Better yet.)

5.1.4 Exercises

Spot, and label specifically, both the template used to prove an existence result (statement of candidate, ...) and where an existence result gives an object that (perhaps modified) becomes that candidate. Exactly where is the output of the existence result used?

5.9: The number 2 has a positive square root.

Proof. Consider f defined by $f(x) = x^2$. Note $f(0) = 0$ and $f(3) = 9$. Also, f is continuous on $[0, 3]$. Since 2 is in $[0, 9]$, there exists c in $(0, 3)$ so that $f(c) = 2$. But then $c^2 = 2$ by the definition of f, and clearly $c > 0$ since it is in the interval $(0, 3)$. So c is a positive square root of 2.

5.10: If f is continuous on $[a, b]$, then f is bounded above on $[a, b]$.

[3]Assume this for now, after drawing pictures, including f and g together.
[4]We can't yet show that our candidate does what it should, but it's right.

Proof. By the Maximum Theorem, f has a maximum point c in $[a, b]$. Let $M = f(c)$. We claim that M is an upper bound for f on $[a, b]$. To show this, let x be any element of $[a, b]$. Then since c is a maximum point, $M = f(c) \geq f(x)$, as required. Thus M is an upper bound for f on $[a, b]$.

(Warning: curves ahead.)

5.11: Suppose f is some function with a limit at 2. Define $-f$ to be the function given by $(-f)(x) = -(f(x))$ for all x. Then $-f$ has a limit at 2.

Proof. We want some number satisfying the definition of limit for $-f$ at the point 2. Let $L = \lim_{x \to 2} f(x)$. We claim $-L = \lim_{x \to 2} -f(x)$. We need to show that for every $\epsilon > 0$ there exists $\delta > 0$ so that if $0 < |x - 2| < \delta$ then we have $|(-f)(x) - (-L)| < \epsilon$.

So let $\epsilon_0 > 0$ be arbitrary.[5] Since f has limit L at 2, we know that for $\epsilon_0 > 0$ there exists some $\delta_* > 0$ so that if $0 < |x - 2| < \delta$ then $|f(x) - L| < \epsilon_0$. We propose δ_* as a suitable δ to accompany ϵ_0 for $-f$. To show δ_* works, note $\delta_* > 0$. Now let x be such that $0 < |x - 2| < \delta_*$. Then $|(-f)(x) - (-L)| = |-(f(x)) - (-L)| = |-1 \cdot (f(x) - L)| = |f(x) - L| < \epsilon_0$ by the choice of δ_* for f. Since x was arbitrary, this holds for all x satisfying $0 < |x - 2| < \delta_*$, so δ_* is adequate for ϵ_0. Since $\epsilon_0 > 0$ was arbitrary, we are done.

5.1.5   *Universal Template*

Wouldn't a template for all proofs (a "universal" template) be nice. But really we'll get a template for proofs involving a universal quantifier ('$\forall$'). It will help that you've used it unconsciously before.

Think back to the geometry proof "The diagonals of a square bisect each other." This is really a universally quantified statement (see Section 3.2), since it could be rephrased with explicit quantifier: "For any square, the diagonals of that square bisect each other." How *did* you cope with the quantifier back then?

Almost certainly you drew a picture of a square with its diagonals and argued from it. You weren't allowed, say, to assume the square had side length 3, since it had to be "any" square. Justification for the approach (if any) might have been: "Yes, you didn't draw all the squares in the world. But your square had nothing special about it (sometimes texts code this by writing "generic") and so what you did for this square you could have done for any square, hence all squares."

On a practical level you avoided mistakes. For example, to prove something about '$\forall T$' where T is an isosceles triangle, your "sample" T is assumed isosceles, but not equilateral, nor an isosceles triangle of side lengths 2, 2, and 3, since neither of these is "generic" isosceles.

---

[5] Note: $\epsilon_0 > 0$ is for a "$-f$" template, not an "f" one.

This is a pretty good description of the template for dealing with any universally quantified thing to prove. To prove something about '$\forall f$,' you start with "let f be arbitrary" and continue on that basis assuming only the given hypotheses for f. Prove what you should with f now viewed as a fixed object (you didn't switch squares in mid proof), and finish by claiming that since you did it for this sample f, the result holds for all f (often this is coded as "since f was arbitrary ..."). The process is this:

i) Consider some "arbitrary" (sample) f of the right type;

ii) Prove what is needed for f viewed as a fixed object;

iii) Remark that since f was arbitrary the result holds for all objects of this type.

Here's an example. Consider f defined by $f(x) = x^2$, the limit of f at $x = 3$, and $\epsilon = .1$; we chose somehow $\delta = \frac{1}{7}$, and now must show that δ "works."[6] We must show that

$$\forall x(0 < |x - 3| < \frac{.1}{7} \Rightarrow |x^2 - 9| < .1).$$

Since the quantified object is "x" we pick an "arbitrary" x such that $0 < |x - 3| < \frac{1}{7}$; call it x_0 so it is clearly "fixed." We want to show $|x_0^2 - 9| < .1$. If successful, since x_0 was arbitrary we conclude by our template that $(0 < |x - 3| < \frac{1}{7} \Rightarrow |x^2 - 9| < .1)$ holds for all x.

Understanding the goals is vital, the proof technicalities much less important. Nonetheless, here we go. We first show that $|x_0 + 3| < 7$.[7] Since $|x_0 - 3| < \frac{1}{7}$, surely $|x_0 - 3| < 1$ and so $x_0 - 3 < 1$, yielding $x_0 < 4$. Also, $-1 < x_0 - 3$, yielding $x_0 > 2$. Clearly, then, $5 < x_0 + 3 < 7$, so $|x_0 + 3| < 7$, as desired.

Now consider $|x_0^2 - 9|$. We have $|x_0^2 - 9| = |(x_0 - 3) \cdot (x_0 + 3)| = |x_0 - 3| \cdot |x_0 + 3| < |x_0 - 3| \cdot 7$ by what was just proved. And so $|x_0^2 - 9| < |x_0 - 3| \cdot 7 < \delta \cdot 7 = \frac{1}{7} \cdot 7 = .1$, yielding $|x_0^2 - 9| < .1$. Since x_0 was arbitrary, we have the result for all x, yielding $\forall x(0 < |x - 3| < \frac{1}{7} \Rightarrow |x^2 - 9| < .1)$.

It is crucial to understand that all the above technical details constitute step *ii)* of the template for proving things with '$\forall$.' Whether technically hard or easy, it's all just that step. We mostly stick with easy functions, but for a fuller repeat of this proof see Exercise 5.22. For now, concentrate on the template.

It is now time for you to do something. Suppose you are faced with f defined by $f(x) = 4x$, the point $x = 3$, and $\epsilon = .1$, and are presented with the candidate $\delta = \frac{1}{4}$. What must you show to prove that δ is satisfactory?

---

[6]Ignore temporarily steps that belong to the template for existence proofs.
[7]This is an unexpected start; it is showing the boundedness of a term.

5.12:

Now go ahead and prove it. You may use the proof above as a model, but realize that this problem is easier, with no "boundedness" required.

5.13:

Repeat the above, but with $\epsilon = .01$ and $\delta = \frac{.01}{4}$ instead.

5.14:

Sincere congratulations: this was a solid first step toward limit proofs.

Aside

It is worth commenting on the language by which the use of the template for proving universally quantified things is cued to the reader. It is terrible. Often used is "Let f be arbitrary ..." and this is an unusual and misleading use of "arbitrary." An arbitrary choice is usually one I have control over, not at all what is meant here: someone else makes an arbitrary choice and I am stuck with it. Some texts introduce the word "generic" instead, which has the advantage that it captures the "no distinguishing features" idea but with the disadvantage that no one in the real world uses that word.

The word "sample" would capture something of the point, because the function we select stands for all of the functions we could have chosen. But "sample" brings to mind statistics, and a statistical sample is usually not *representative* of the population in question but only *random*. With a sample of voters, we'd like to get a sample exactly mimicking the whole population of voters (e.g., our sample 47% pro-wombat if the population is 47% pro-wombat), and generally this doesn't happen. We take a random sample, knowing that were we to do so over and over we have a certain (good) probability of being close to the population figure. But since in proving something for all functions f we aren't allowed only to be "close" or "probably right," this won't do.

Another possible word is "representative": grab a function which somehow captures in it all of the quirks and difficulties of all the functions we could have picked. But there is no single such function, nor can we pick the "worst function of all" partly because there probably is no such function (can one function display all the types of failure of continuity?). Also, we are responsible for all functions, not just the bad ones. So this word isn't right either.

Conclusion: you must recognize what the language is cueing not because the customary language is perfect but because it is standard. The cueing in

English is somehow clumsy, but the underlying logic is fine, and it's vital to know what's going on.
End Aside

5.1.6 Exercises

5.15: With $f(x) = -4x$, $x = 3$, $\epsilon = .1$, and $\delta = \frac{.1}{4}$, prove carefully that δ works. Repeat with $\epsilon = .05$ and $\delta = \frac{.05}{4}$.

5.16: Consider $f(x) = 6x$, $x = 10$, $\epsilon = .1$, and $\delta = \frac{.1}{6}$. Prove carefully that δ works. Repeat with $\epsilon = .05$ and $\delta = \frac{.05}{6}$.

5.17: Consider $f(x) = x^2$, $x = 0$, $\epsilon = .1$, and $\delta = .1$. Prove δ works. Repeat with $\epsilon = .05$ and $\delta = .05$. Do larger δ's work for these ϵ's? Proof?

5.18: With f defined by $f(x) = x \cdot \sin \frac{1}{x}$, $x = 0$, the proposed limit 0, $\epsilon = .1$, and $\delta = .1$, prove δ works, carefully. Repeat with $\epsilon = .01$ and $\delta = .01$.

5.19: Consider the function f defined by

$$f(x) = \begin{cases} x, & x \text{ rational,} \\ 0, & x \text{ irrational.} \end{cases}$$

Faced with $x = 0$, $\epsilon = .1$ and $\delta = .1$, prove this δ is satisfactory.

5.20: We digress to see how to show a universally quantified thing is false. To show a statement with a '$\forall$' quantifier is false is to find an object in the domain of the quantified variable yielding a false statement when it is inserted. So $\forall T(T$ is a triangle $\Rightarrow T$ is equilateral) is false since the particular triangle T_0 which is the $3, 4, 5$ right triangle makes the implication false.

With this in mind, consider the function f defined by

$$f(x) = \begin{cases} 1, & x > 0, \\ 0, & x = 0, \\ -1, & x < 0, \end{cases}$$

the proposed limit 0, $\epsilon = .1$, and $\delta = .1$. Show this δ does *not* work by showing $\forall x(0 < |x - 0| < .1 \Rightarrow |f(x) - 0| < .1)$ is false. (See Exercise 1.43 for help.)

Note: although it is true, we *haven't* shown that $\lim_{x \to 0} f(x) \neq 0$, but only that for this particular ϵ the proposed δ fails. Conceivably a more clever choice of δ would work.

5.21: Again we show the '$\forall(x)$' form in a limit may be false.

Consider $f(x) = \sin \frac{1}{x}$, $x = 0$, the proposed limit 0, $\epsilon = .1$ and $\delta = .1$. Show this δ fails (Section 1.5 may help). Again, we haven't shown $\lim_{x \to 0} f(x) \neq 0$, much as in the last problem.

5.22: Return to f defined by $f(x) = x^2$, the limit of f at $x = 3$, $\epsilon = .1$, $\delta = \frac{1}{7}$ proposed, and our try to show δ "works." We'll show the private scratchwork to find δ, which will motivate the odd $|x_0 + 3| < 7$ step.

We must show $\forall x (0 < |x - 3| < .1/7 \Rightarrow |x^2 - 9| < .1)$. Via the '$\forall$' template, we will really need, for some x_0 arbitrary, $|x_0^2 - 9| < .1$ assuming $|x_0 - 3| < \delta$. Our sole power is to choose δ cleverly. Well, $|x_0^2 - 9| = |(x_0 - 3) \cdot (x_0 + 3)| = |x_0 - 3| \cdot |x_0 + 3|$,[8] to be forced less than .1. How are we doing? One of these, $|x_0 - 3|$, is a thing we can make as small as we like by choosing δ, so if it were all $\delta = .1$ would work. But this factor gets multiplied by $|x_0 + 3|$, not necessarily small.

Here's the trick. Yes, $|x_0 + 3|$ need not be small, but we expect x_0 to be close to 3. If so, $|x_0 + 3|$ should be close to 6. Since x_0 could be larger than 3, $|x_0 + 3| \leq 6$ isn't true. But if x_0 is quite close to 3, surely $|x_0 + 3|$ will be less than 7. Then $|x_0^2 - 9| = |(x_0 - 3) \cdot (x_0 + 3)| = |x_0 - 3| \cdot |x_0 + 3| < \delta \cdot 7$, so $\delta \cdot 7 < .1$ works. Right, $\delta = \frac{1}{7}$, to make $|x_0 - 3|$ so "extra" small that even when multiplied by 7 the product will be less than .1. It's almost a "worst case" analysis to see how bad the term $|x + 3|$ could be, and bounded at its worst by 7.

OK, end scratchwork: $\delta = \frac{1}{7}$ is our candidate, the rest of proof as written. This decomposition of our product into a small term and a "bounded" term is extremely useful. (See, in fact, Exercise 5.18, where $\sin \frac{1}{x} \leq 1$.)

Imitate this scratchwork, choice of δ, and proof via bounding, for $f(x) = x^2$, $x = 4$, the correct limit, and $\epsilon = .1$.

5.23: For f defined by $f(x) = 4x^2$, $x = 2$, the correct limit, and $\epsilon = .1$, find a suitable δ by scratchwork, and show it works.

5.24: Repeat for $f(x) = 4x^2 \sin x \cos x$, $x = 0$, the correct limit, and $\epsilon = .1$.

5.2 Limits, Finally

Bad news: in spite of all the work above, we have not done a limit proof yet, since while we grappled carefully with the '$\forall x$' part, we did not cope with '$\forall \epsilon > 0$' but only with $\epsilon = .1$ (or .01, etc.). Look again at the definition of limit:

Definition 5.2.1 *Let f be a function defined on an open interval containing the point b, except possibly at b itself. We say $\lim_{x \to b} f(x) = L$ if*

$$\forall(\epsilon > 0)(\exists(\delta > 0)(\forall(x)(0 < |x - b| < \delta \Rightarrow |f(x) - L| < \epsilon))).$$

---

[8]This works forward from what we ought to prove, an approach illegal in a proof, but anything goes in private scratchwork.

This contains *three* quantifiers, the '$\forall$' quantifying ϵ, '$\exists$' quantifying δ, and '$\forall$' quantifying x. The following shows what we need:

$$\forall(\epsilon > 0)(\exists(\delta > 0)(\underbrace{\forall(x)(0 < |x - b| < \delta \Rightarrow |f(x) - L| < \epsilon)}_{\forall x})).$$

with $\exists \delta$ and $\forall \epsilon$ underbraces.

The quantifier nesting is annoying, but it is exactly mirrored by template nesting within the proof. (Recall the subscript 0 signals the fixed "arbitrary" thing in '$\forall$' proofs.) We work our way from the outside in, and so first use the template for '$\forall \epsilon$':

1. Let $\epsilon_0 > 0$ be arbitrary.

2. Show what we need to show for that particular ϵ_0.

3. From this, the result holds for all $\epsilon > 0$.

For the particular $\epsilon_0 > 0$, we must show there exists a $\delta > 0$ such that something happens (the "something" being '$\forall(x)(0 < |x - a| < \delta \Rightarrow |f(x) - L| < \epsilon_0$'). Call our candidate (when found) δ_*, and inserting the existence template in the above outline we get:

1. Let $\epsilon_0 > 0$ be arbitrary.

2. Show what we need to show for that particular ϵ_0:

   i) $(*)$ Find somehow a candidate for the good δ_*;

   ii) Commit ourselves publicly that this is the candidate;

   iii) Prove δ_* satisfies $\forall(x)(0 < |x - b| < \delta_* \Rightarrow |f(x) - L| < \epsilon_0)$.

3. From this, the result holds for all $\epsilon > 0$.

Finally, supposing δ_* found, to complete step *iii)* we use the template for '$\forall x$'. Inserting this template properly yields

1. Let $\epsilon_0 > 0$ be arbitrary.

2. Show what we need to show for that particular ϵ_0:

   i) $(*)$ Find somehow a candidate for the good δ_*;

   ii) Commit ourselves publicly that this is the candidate;

   iii) Prove δ_* satisfies $\forall(x)(0 < |x - b| < \delta_* \Rightarrow |f(x) - L| < \epsilon_0)$:

      (a) Let x_0 such that $0 < |x_0 - b| < \delta_*$ be arbitrary.

      (b) Show what we need to show for that particular x_0, namely $|f(x_0) - L| < \epsilon_0$.

(c) From this, the result holds for all x.

3. From this, the result holds for all $\epsilon > 0$.

Crucial Point I

Some good news is that when we are working with x, say, we need not worry about the '$\forall \epsilon > 0$' clause; the template for '$\forall \epsilon$' leaves us with ϵ_0 fixed. And at the δ level, we aren't worrying about either the '$\forall \epsilon > 0$' clause or the '$\forall x$' clause. And in the very inside, ϵ_0, δ_*, and x_0 are all fixed and all of the quantifiers are (temporarily) gone.

Crucial Point II

Our prior δ candidates were numerical ($\frac{1}{4}$ or something) given numerical values for ϵ (e.g., .1). Occasionally we got a formula for δ in terms of ϵ (see, for example, Exercises 1.21 and 1.22). Here we face simply ϵ_0, and so such a "formula" for our δ will be needed. Except in very rare and trivial cases, a numerical value for δ_* won't work since δ_* should vary with ϵ_0.

We need an example, so return to $f(x) = 4x$ and $x = 3$ (from Section 1.3). For $\lim_{x \to 2} f(x) = 12$, we need

$$\forall(\epsilon > 0)(\exists(\delta > 0)(\forall(x)(0 < |x - 3| < \delta \Rightarrow |4x - 12| < \epsilon))).$$

Hold tight to the outline above; here we go.

Let $\epsilon_0 > 0$ be arbitrary. What we need for ϵ_0 is that there exists a δ, so we have to find our candidate δ_*. Scratchwork, please (Section 1.3):

Scratchwork

We want $|4x - 12| < \epsilon_0$, and we get to assume $|x - 3| < \delta_*$. Now $|4x - 12| = 4 \cdot |x - 3| < \epsilon_0$ goes with $|x - 3| < \epsilon_0/4$, so $\delta_* = \epsilon_0/4$ seems to be indicated. (Cheat slightly: in Section 1.3 this worked for $\epsilon_0 = .1$, good.) So try $\delta_* = \epsilon/4$.

End Scratchwork

Proof. Let $\epsilon_0 > 0$ be arbitrary. Take $\delta_* = \epsilon_0/4$, and we must show $\forall(x)(0 < |x - 3| < \delta_* \Rightarrow |4x - 12| < \epsilon_0)$. With our δ_*, this is $\forall(x)(0 < |x - 3| < \epsilon_0/4 \Rightarrow |4x - 12| < \epsilon_0)$. For this, let x_0 be arbitrary so that $0 < |x_0 - 3| < \epsilon_0/4$. Then $4 \cdot |x_0 - 3| < \epsilon_0$, so $|4x_0 - 12| < \epsilon_0$ as desired. (Everything to follow is digging our way out of the three quantifier templates, in the order "x", "δ", "ϵ".) Since x_0 was arbitrary, the result holds for all x. Since the result holds for all x, the δ_* we found was successful. Since $\epsilon_0 > 0$ was arbitrary, and we found a successful δ for it, there is a δ for every $\epsilon > 0$, what we had to prove. Done.

Take some time to absorb this.

5.25:

This is a typical limit problem in outline. Other problems may have harder δ-finding scratchwork, or harder δ-validation in the proof. But after you have done the exercises below, you may truly say that you have done some limit proofs (sometimes called ϵ-δ proofs) in full detail.

5.2.1 Exercises

5.26: Consider f defined by $f(x) = -4x$ and $x = 3$. Prove that the limit is what you think it is. (Exercise 5.15 may help.)

5.27: Repeat with $f(x) = 6x$ and $x = 10$. (Cf. Exercise 5.16.)

5.28: Repeat with $f(x) = x^2$ and $x = 0$. Assume temporarily that "all ϵ so that $1 > \epsilon > 0 \ldots$" is enough, and cf. Exercise 5.17.

5.29: Repeat, including the $\epsilon < 1$ assumption, with $f(x) = x \cdot \sin \frac{1}{x}$ and $x = 0$. (Exercise 5.18 may be useful.)

5.30: Repeat with f (cf. Exercise 5.19) defined by

$$f(x) = \begin{cases} x, & x \text{ rational,} \\ 0, & x \text{ irrational.} \end{cases}$$

5.2.2 Nasty Technical Details

We present here the (ugly) scratchwork for finding δ given ϵ for x^2 at $x \neq 0$, and see in passing a useful trick for the "large ϵ" problem assumed away in Exercises 5.28 and following. Finally, we turn to showing that a limit does not exist.

We've considered the square function at an x not 0, namely $x = 3$. Even for $\epsilon = .1$ we worked extra hard to get $|x^2 - 9|$ under control: write it as $|x - 3| \cdot |x + 3|$ and "bound" the $|x + 3|$ term (by 7) to get $\delta = \frac{1}{7}$ to work. But what about general ϵ_0, not just .1?

Trying $\delta_* = \epsilon_0/7$ is reasonable, doesn't quite work, but can be improved to work. With that δ_*, we must show $\forall x (0 < |x - 3| < \epsilon/7 \Rightarrow |x^2 - 9| < \epsilon_0)$. So let x_0 such that $0 < |x_0 - 3| < \epsilon_0/7$ be arbitrary and start out $|x_0^2 - 9| = |x_0 - 3| \cdot |x_0 + 3| < \frac{\epsilon_0}{7} \cdot |x_0 + 3|$. Stuck: we want to say that x_0 is pretty close to 3 but can't quite. Suppose that some troublemaker handed us $\epsilon_0 = 100$. Then $|x_0 - 3| < \frac{\epsilon_0}{7}$ would guarantee only that

$$\frac{-100}{7} < x_0 - 3 < \frac{100}{7}$$

or

$$\frac{-79}{7} < x_0 < \frac{121}{7}.$$

So $x_0 + 3$ might be, say, 20 (not 7), and the above effort at a proof fails.

There's a trick. We need not take $\delta_* = \frac{\epsilon}{7}$ if ϵ is large; in fact, we insist on $\delta_* \le 1$, no matter what else. Technically, we let $\delta_* = \min(\epsilon_0/7, 1)$. Then $|x_0 + 3| < 7$ follows from $|x_0 - 3| < \delta_*$, as before. This also removes the $\epsilon < 1$ assumptions in previous Exercises; allow any ϵ at all, but make sure $\delta_* \le 1$.

Prove that $\lim_{x \to 3} x^2 = 9$ using this trick.

5.31:

There is a standard mistake to avoid. For the problem above, someone will surely propose that we take $\delta_* = \epsilon_0/|x+3|$, or perhaps $\delta_* = \epsilon_0/|x_0+3|$. It's reasonable, natural, and completely wrong.

In the first formulation δ_* isn't a number depending on ϵ, because of the x. What is x? Given $\epsilon = .1$, δ_* should come from your formula (as the Section 5.2 formula yielded for $\epsilon = .1$ what we had already obtained in Section 5.1.5). Since it doesn't, this isn't a legal value of δ. A δ so defined can slide around depending on x, when it should be fixed.

For $\delta_* = \epsilon_0/|x_0+3|$ x_0 *is* some fixed number, so the above objection isn't it. Problem: δ is proposed at a level where ϵ_0 is fixed, but we aren't even at the level of x yet, so x_0 hasn't occurred. Our value for δ may (usually will) depend on ϵ_0 (ϵ_0 has already been introduced) but may not depend on x or x_0, not yet introduced.

We could make all this precise with more logical tools, but we won't. A rule of thumb to get you through is that your δ will be wrong if it contains x or x_0 in it. Indeed, the "bounding" techniques actually replace the x_0 you might want to use with some sort of "worst case" x_0. That 7 in the denominator of δ_* is $|x_0+3|$ with the worst case x_0, namely $x_0 = 4$, inserted. Similarly, if you wanted to try $\delta_* = \epsilon_0/|x_0^2 + 3|$ in a problem, you'll win by showing that x_0 will be in $[2, 5]$, so $x^2 < 25$, so $|x_0^2 + 3| < 28$; then try $\delta_* = \epsilon_0/28$.

Finally, how do you show that $\lim_{x \to b} f(x) = L$ fails? Technically, we must show that *there exists* $\epsilon > 0$ such that *for all* $\delta > 0$ *there exists* x such that $0 < |x - a| < \delta$ but $|f(x) - L|$ is not less than ϵ. We simply record here that our approach in Sections 1.4 and 1.5 was correct: find an $\epsilon > 0$ with no satisfactory δ. To show a proposed δ fails we must find x (in our δ region) so that $f(x)$ is bad.

5.2.3 Exercises

5.32: Prove that $\lim_{x \to 4} x^2$, and $\lim_{x \to -4} x^2$, are what you think.

5.33: Prove $\lim_{x \to 0} x^3$ is what you think.

5.34: Prove that $\lim_{x \to 2} x^3 = 8$. Hint: $x^3 - a^3 = (x - a)(x^2 + ax + a^2)$.

5.35: Consider the function f defined by

$$f(x) = \begin{cases} 1, & x > 0, \\ 0, & x = 0, \\ -1, & x < 0. \end{cases}$$

Argue that $\lim_{x \to 0} f(x) \neq 1$, and $\lim_{x \to 0} f(x) \neq 0$. Hint: $\epsilon = .1$ will work just fine.

5.36: Continue with the previous function. Assuming that that limit is not $1, -1$, or 0, what remains to show that the limit does not exist at 0? For a typical example of what needs to be done, make the argument.

5.37: Consider f defined by $f(x) = \sin(\frac{1}{x})$ for $x \neq 0$, and define it any way you like at $x = 0$. Argue that $\lim_{x \to 0} f(x) \neq 0$. Argue that for any L such that $L > 1$, $\lim_{x \to 0} f(x) \neq L$. (You will indeed use the template for '$\forall L$.') What about L's such that $L < -1$? L in the range $(0, 1]$? Can you finish?

5.38: Consider the following (a standard for counterexamples):

$$f(x) = \begin{cases} 1, & x \text{ rational}, \\ 0, & x \text{ irrational}. \end{cases}$$

Show that $\lim_{x \to 0} f(x)$ is neither 0 nor 1. Useful fact: any open interval contains both a rational number and an irrational number.

5.39: Here's another variant of one of the functions above: define f by

$$f(x) = \begin{cases} x^2 \cdot \sin(\frac{1}{x}), & x \neq 0, \\ 0, & x = 0. \end{cases}$$

Does it have a limit at 0? If it does, prove it. If not, prove that.

5.40: Repeat with f defined by

$$f(x) = \begin{cases} x, & x \text{ rational}, \\ 0, & x \text{ irrational}. \end{cases}$$

5.41: Repeat with f defined by

$$f(x) = \begin{cases} \frac{1}{x}, & x = 1/2^n, \text{ some } n = 1, 2, \ldots, \\ 0, & \text{otherwise}. \end{cases}$$

6

Limit Theorems

The theorems in this chapter might be generally classified as "new limits from old" because many of them have the flavor "if f has a limit at b (or f and g have limits at b) then some variant of f (or combination of f and g) has a limit at b, in fact, such-and-such limit." But for such theorems we need an improvement in our technique, to handle limit proofs not for a specific function ("formula") at a particular point.

Here's a familiar example of this new type of theorem.

Theorem 6.0.2 *Suppose f and g are functions satisfying $\lim_{x \to b} f(x) = L$ and $\lim_{x \to b} g(x) = M$. Then $f + g$ satisfies $\lim_{x \to b}(f + g)(x) = L + M$.*

Recall "$f + g$" is defined by $(f + g)(x) = f(x) + g(x)$ for all x, so, equivalently, this theorem states that, if the individual limits exist,

$$\lim_{x \to b} f(x) + g(x) = \lim_{x \to b} f(x) + \lim_{x \to b} g(x).$$

The limit of the sum is the sum of the limits: what else?

But the proof isn't so easy. First, there are implicit universal quantifiers on f and g, each of which has a limit at b. So, via the template, we pick an arbitrary f with limit at b and an arbitrary g with limit at b. So an outline of the proof will look like

> Pf. Let f_0 and g_0 be arbitrary functions with limits L and M at b, respectively.
>
> . . .
>
> Thus, $\forall(\epsilon > 0)\exists(\delta > 0)(\forall(x)(0 < |x - a| < \delta \Rightarrow |(f_0 + g_0)(x) - (L + M)| < \epsilon))$.

So $\lim_{x \to b} f_0(x) + g_0(x) = L + M$ as desired.

Since f_0 and g_0 were arbitrary functions, the result holds in general. ∎

All our work is confined to the box, but we have f_0 and g_0 instead of a single specific function; that f_0 and g_0 are "fixed" in the interior of the proof isn't as good as formulas.

It's actually worse. Expand the outline a little to include our "for all ϵ" template and also our announcement ("exists" template) of a δ candidate:

Pf. Let f_0 and g_0 be arbitrary functions with limits L and M at b, respectively.

> Let $\epsilon_0 > 0$ be arbitrary.
> Let $\delta_* =$ **(inspiration needed here!)**.
> . . .
> Therefore we have found a δ_* to go with ϵ_0 such that
> $\forall(x)(0 < |x - a| < \delta_* \Rightarrow |(f_0 + g_0)(x) - (L + M)| < \epsilon_0)$.

So for ϵ_0, there does exist a $\delta_* > 0$ such that $\forall(x)(0 < |x-a| < \delta_* \Rightarrow |(f_0 + g_0)(x) - (L + M)| < \epsilon_0)$. Thus, since $\epsilon_0 > 0$ was arbitrary, $\forall(\epsilon > 0)\exists(\delta > 0)(\forall(x)(0 < |x - a| < \delta \Rightarrow |(f_0 + g_0)(x) - (L + M)| < \epsilon))$.

So $\lim_{x \to b} f_0(x) + g_0(x) = L + M$ as desired.

Since f_0 and g_0 were arbitrary functions, the result holds in general. ∎

The problem is at, or rather before, the "inspiration needed here" marker. To find δ_* in the past we did scratchwork with the specific function (again, a *formula*) to get a candidate. But there is no formula for f_0 or g_0 to use to "compute" a formula for δ_* in terms of ϵ_0 (let alone two functions instead of one). Conclusion: we need some new source for a δ_*.

One suspects δ for $f + g$ has something to do with f and something to do with g. What do we know? Since $\lim_{x \to b} f(x) = L$, for any $\epsilon > 0$ there is a δ_f to go with f such that something happens for some x's and f. Similarly, with any $\epsilon > 0$ there is δ_g to go with g. That is, we have two *existence* results from the facts about f and g and we need a candidate for an existence proof.

The language hints at the process in Section 5.1.3, exactly the use of existence hypotheses to produce candidates for an existence proof. Review that section thoroughly.

6.1:

We will use that approach: use the δ_f production ability of f and the δ_g production ability of g to produce δ_* for $f + g$. Here's an analogy to this sum theorem. Suppose you assemble widgets from parts A and B obtained from other suppliers. If you must ensure that A and B glued together yield a total length of 10 inches $\pm.01$ inch, you don't control errors in A and B. But if A's producer can guarantee as small an error as you need, and the same for part B, you can control the error of the total as closely as you need. You don't compute δ_A or δ_B, but since those controls are available you can set the δ of the combination to control the total error.

We'll work up to this result after some simpler ones, but all with existence results producing existence candidates.

6.1   A Single δ

In our first example we search for a δ with only one existence hypothesis to use. This gives only one "machine" which will produce a δ upon being handed an ϵ (so no "combining" two δ's to get a candidate . . . yet). Consider the following Proposition (seen first in Section 1.5).

Proposition 6.1.1 *Suppose f and g are functions each defined on all of $\mathbf{R}$ except possibly a point b, and such that for each point x of $\mathbf{R}$, except possibly b, we have $f(x) = g(x)$. Then if $\lim_{x \to b} f(x)$ exists, so does $\lim_{x \to b} g(x)$ and they are equal.*

If one accepts the intuition that the limit has nothing to do with the value of the function at the point, and that it is a local property depending only on values near the point, this result is obvious, because in the crucial region (near b but not at b) there is no difference between f and g. As a matter of fact, support that intuition graphically before we begin the proof discovery.

6.2:

Proof Discovery
   (**Vital**: make sure the outline from Section 5.2 is in front of you.)
   Suppose that $\lim_{x \to b} f(x)$ exists, and call it L. One would suspect, then, that $\lim_{x \to b} g(x) = L$. By definition, this is

(6.1) $\forall(\epsilon > 0)\exists(\delta > 0)(\forall(x)(0 < |x - b| < \delta \Rightarrow |g(x) - L| < \epsilon)).$

So we'll start with $\epsilon_0 > 0$ arbitrary (using the '$\forall\epsilon$' template). We need a candidate δ_* for δ. Where is it to come from?
   Since $\lim_{x \to b} f(x) = L$, $\forall(\epsilon > 0)\exists(\delta > 0)(\forall(x)(0 < |x - b| < \delta \Rightarrow |f(x) - L| < \epsilon))$. View this as a machine which, when handed an ϵ, will hand us a δ.[1] Note that we have an ϵ in our problem, namely ϵ_0. Why not

---

[1] Of course, there are no guarantees for non-approved uses: the δ we are handed is one useful for f, and checking that it works for g is our problem. With no other

insert ϵ_0 into the δ producing machine associated with f and get a δ_*? Let's do so.

Crucial: this δ_* is a possible, but not guaranteed, candidate for the δ we need to complete the proof of (6.1). Checking it consists of showing that

$$\forall(x)(0 < |x - b| < \delta_* \Rightarrow |g(x) - L| < \epsilon_0).$$

As usual we let x_0 be arbitrary so that $0 < |x_0 - b| < \delta_*$, and try for $|g(x_0) - L| < \epsilon_0$, then getting the claim for all x. Distinguish carefully between what we know about x_0 and what we want.

Know

Since δ_* was produced by inserting ϵ_0 into the definition of $\lim_{x \to b} f(x) = L$, we know that

$$\forall(x)(0 < |x - b| < \delta_* \Rightarrow |f(x) - L| < \epsilon_0).$$

Since x_0 *is* an x such that $0 < |x - b| < \delta_*$, we may therefore deduce that

(6.2) $|f(x_0) - L| < \epsilon_0.$

Note this is, as might be expected, a statement about f.

Want

What we would like is

(6.3) $|g(x_0) - L| < \epsilon_0.$

(We want this because it would complete what we need for x_0, and then we'll finish up with "since x_0 was arbitrary ...".) Well?

6.3:

In general (6.2) and (6.3) are unrelated: one is about f, the other about g. But f and g are related, since $f(x) = g(x)$ for many x. Might x_0 be such an x?

6.4:

Since $|x_0 - b| > 0$, we have $x_0 \neq b$. So $f(x_0) = g(x_0)$. And so (6.2) gives (6.3): key idea in place.

Usually during or after proof discovery you spot missing things. Note that this is really a theorem about all functions f and g, so the '$\forall f$' and '$\forall g$' templates are needed, but other than that we are in good shape.

―――――――――

candidate in sight, however, we press on.

Write the proof yourself: you have an outline and the key idea. Take some time to digest when you're done.
Proof

6.5:

We can improve the result. We have assumed that f and g agree everywhere except at b, but we really only need that they agree in an open interval around b (except at b). Draw a graph to convince yourself this is enough; then modify the proof (using a good idea we've seen) to cover this case.

6.6:

6.1.1 Exercises

6.7: Let f be a function with a limit at the point b. Prove that $-f$ has a limit at b, where $-f$ is defined by $(-f)(x) = -(f(x))$ for all x.

6.8: Suppose $\lim_{x \to 3} f(x) = L$. Prove that $\lim_{x \to 3}(f(x) + 5) = L + 5$ from the definition of limit. Modify to cover all constants d, not just 5.

6.1.2 A Single δ, but Worse

Even the case of a single "source" for candidate δ's can be worse than what is shown above. We can illustrate the difficulty, and its resolution, by considering the following (Exercise 6.7 was a special case).

Proposition 6.1.2 *Let f be a function with limit L at the point b. Then if c is any constant, cf has limit cL at b, where cf is the function defined by $(cf)(x) = c \cdot f(x)$ for all x.*

Concrete example: if we know $\lim_{x \to 3} x^2 = 9$, then $\lim_{x \to 3} 4x^2 = 36$.

Discover the key idea of the proof and then write it up. Ignore the case $c = 0$, for although it is separate, it is easy.

6.9:

Welcome back from your efforts. We'll start out with a natural attempt, which, since it is doomed, we want to warn you against.

Proof Discovery: Attempt 1
We must show that $\forall(x)(0 < |x - b| < \delta_* \Rightarrow |c \cdot f(x) - c \cdot L| < \epsilon_0)$ (starting with $\epsilon_0 > 0$, what else?). We need a candidate for δ_*. We know

that $\forall(\epsilon > 0)\exists(\delta > 0)(\forall(x)(0 < |x - a| < \delta \Rightarrow |f(x) - L| < \epsilon_0))$. This is a δ generating machine to use, so why not hand it ϵ_0? Do so, and we produce δ_*. Now this δ_* doesn't take into account the c that we are going to be multiplying f by, so we had better use δ_*/c instead.

We interrupt your regularly scheduled programming to announce that we are in deep trouble. A minor reason is the possibility $c < 0$ (see Exercise 6.7), but this is fixed by $\delta_*/|c|$. It's much worse than that.

Let's start a very careful and suspicious attempt to prove that $\delta_*/|c|$ works. We must show $(\forall(x)(0 < |x-b| < \delta_*/|c| \Rightarrow |c \cdot f(x) - cL| < \epsilon_0)$. So let x_0 be arbitrary such that $0 < |x_0 - b| < \delta_*/|c|$. We want $|c \cdot f(x) - cL| < \epsilon_0$. This is the same as

(6.4) $$|f(x) - L| < \epsilon_0/|c|.$$

But we cannot conclude this on the basis of what δ_* does for f.

If $0 < |x - b| < \delta_*$ we have $|f(x) - L| < \epsilon_0$. Suppose c were 5, ϵ_0 were .1, and δ_* were .001. We know that if x is within .001 of b then $f(x)$ is within .1 of L. We want that if x is 5 times closer to b (within $\frac{.001}{5}$) then $f(x)$ is 5 times closer to L (within $\frac{1}{5}$) — our goal in inequality (6.4). Unfortunately, this doesn't work.

Here's an example. Consider $f(x) = \sqrt[3]{x}$, $b = 0$, $\epsilon_0 = .1$, $\delta_* = .001$. Check via calculator, graphically, and all other ways that δ_* works for ϵ_0.

6.10:

Our try above was that if ϵ were 5 times smaller (thus, $\frac{1}{5} = .02$) then a δ which is 5 times smaller (thus, $\frac{.001}{5} = .0002$) will work. Again, check that this is what are trying to do in inequality (6.4). But check, numerically and graphically, that $\delta = \frac{.001}{5}$ fails.

6.11:

Note that we are *not* saying that *no* value of δ works for $\epsilon_0 = \frac{1}{5}$, but only that "if δ works for ϵ then $\delta/5$ works for $\epsilon/5$" is false.

Return to our goal: to make $f(x)$ so close to L that $c \cdot f(x)$ is still very close to $c \cdot L$. Our attempt to do so by getting $f(x)$ close to L, and then forcing x to be c times closer to b, failed. What's to do?

Proof Discovery: Attempt 2

We want $|f(x) - L| < \epsilon_0/|c|$ (so that $|c \cdot f(x) - c \cdot L| < \epsilon_0$). What can we say about $\epsilon_0/|c|$? It is a (probably small) positive number.

VITAL, CRUCIAL, point

Up until now, when we had an ϵ_0 in our problem, and an ϵ-input δ-producing machine, we have given it ϵ_0. Seems sensible. But we may give it **any** (any, any, *any*) positive number and get a δ out. Inspiration: give it the particular small positive number $\epsilon_0/|c|$.

Formally, we know that $\forall(\epsilon > 0)\exists(\delta > 0)(\forall(x)(0 < |x - a| < \delta \Rightarrow |f(x) - L| < \epsilon_0))$. Set ϵ to $\epsilon_0/|c|$. We get $\delta_2 > 0$ such that $\forall(x)(0 < |x - a| < \delta_2 \Rightarrow |f(x) - L| < \epsilon_0/|c|)$. It works(!):

Proof.

We need $\forall(\epsilon > 0)\exists(\delta > 0)(\forall(x)(0 < |x - b| < \delta \Rightarrow |c \cdot f(x) - c \cdot L| < \epsilon))$. Let $\epsilon_0 > 0$ be arbitrary. We know that $\forall(\epsilon > 0)\exists(\delta > 0)(\forall(x)(0 < |x - b| < \delta \Rightarrow |f(x) - L| < \epsilon))$. In this, set $\epsilon = \epsilon_0/|c|$ and produce δ_2 such that

(6.5) $$\forall(x)(0 < |x - b| < \delta_2 \Rightarrow |f(x) - L| < \epsilon_0/|c|).$$

We show that $\forall(x)(0 < |x - b| < \delta_2 \Rightarrow |c \cdot f(x) - c \cdot L| < \epsilon_0)$. Let x_0 such that $0 < |x_0 - b| < \delta_2$ be arbitrary. From equation (6.5) we have $|f(x_0) - L| < \epsilon_0/|c|$. Multiply on both sides by $|c|$ to get $|c| \cdot |f(x_0) - L| < \epsilon_0$. Passing the $|c|$ inside the absolute values, we get $|c \cdot f(x_0) - c \cdot L| < \epsilon_0$, as desired. Since x_0 was arbitrary, the result holds for all x. Thus δ_2 works for ϵ_0. Since ϵ_0 was arbitrary, the result holds for all $\epsilon > 0$, and thus we have succeeded in showing $\forall(\epsilon > 0)\exists(\delta > 0)(\forall(x)(0 < |x - b| < \delta \Rightarrow |c \cdot f(x) - c \cdot L| < \epsilon))$, and therefore that $\lim_{x \to b} c \cdot f(x) = c \cdot L$ as desired.

Pause and ponder, long and hard, because this idea of inserting something besides ϵ_0 into the δ-producing machine is crucial. Note finally that we are missing a '$\forall f$' template, but that's easy.

6.12:

6.1.3 *Exercises*

6.13: Let f be a function with limit zero at the point b. Let g be any function such that $-5 < g(x) < 3$ for all x, and such that g is defined at all points of the real line. Prove that $\lim_{x \to b} f(x) \cdot g(x) = 0$.

The basic idea in the problems to follow is similar to that above, in that there is the use of something known to exist from one source (often a δ) to produce a candidate for something we need (often another δ). The results are milestones along the way to proving that if $\lim_{x \to b} f(x) = L$, then $\lim_{x \to b} \frac{1}{f(x)} = \frac{1}{L}$. Of course $L = 0$ fails, but worse is the potential problem that $\lim_{x \to b} \frac{1}{f(x)}$ might fail because $1/f$ might be undefined infinitely often near b. (Note that each zero of f, not a problem, turns into an undefined point for $1/f$, a big problem; recall also that $\sin \frac{1}{x}$ is zero often near $x = 0$.) Showing this can't happen is a good deal of the work. The details are technical, and we'll build up slowly.

6.14: Suppose $\lim_{x \to b} f(x) = L$. Show that there is a punctured neighborhood of b on which $f(x) < L + 1$ for all x. [Hint: First, phrase "there is a punctured neighborhood of b ..." as " there exists $\delta > 0$ such that for all x such that $0 < |x - b| < \delta$ ... ." This makes the problem a familiar "looking for a δ" one. Now insert $\epsilon = 1$ into the limit definition. Out comes δ. Is this δ suitable for the δ you are looking for? Do you need to insert a different ϵ?]

6.15: Suppose $\lim_{x \to b} f(x) = L$. Show that there is a punctured neighborhood of b and an M so that $|f(x)| \leq M$ for all x on the punctured neighborhood. [Hint: The previous problem gave you an interval (or equivalently a δ) on which $f(x) < $ (something). Use the same technique to get a δ_2 such that on the appropriate associated interval $f(x) > $ (something else). How do you get an interval on which you have f bounded both above and below? (For example, if $\delta_1 = .1$ guarantees the upper bound 5 and $\delta_2 = .05$ guarantees the lower bound -7; what is some number M such that $|f(x)| \leq M$? What's δ?)]

6.16: Here are some definitions:

Definition 6.1.3 *A function f defined on a set S is <u>bounded below</u> if there exists d so that $f(x) > d$ for all x in S; in this case we say f is bounded below by d.*

The definition of bounded above is completely similar, although M is a standard letter. Now prove the following proposition.

Proposition 6.1.4 *Let f be a function bounded below on a set S by $d > 0$. Then $\frac{1}{f}$ is bounded above on S.*

(Note: you are not looking for a δ from a δ, but an M from a d.

6.17: Suppose $\lim_{x \to b} f(x) = L$, and suppose in addition $L > 0$. A graph (go ahead) will convince you that f is actually positive near b, because values of f close to L can't also be close to 0. We will prove more.

Proposition 6.1.5 *Suppose $\lim_{x \to b} f(x) = L$ and that $L > 0$. Then there exists a punctured neighborhood of b on which f is bounded below by $L/2$.*

[Hints: Rephrase "punctured neighborhood" in terms of δ, and use the following odd but useful trick: insert $L/2$ as ϵ into the definition of limit.]

6.18: Here's the theorem.

Theorem 6.1.6 *Suppose $\lim_{x \to b} f(x) = L$ and further that $L \neq 0$. Then $\lim_{x \to b} \frac{1}{f(x)} = \frac{1}{L}$.*

We'll assume $L > 0$ for convenience. Doing scratchwork, we are faced with showing

$$\left| \frac{1}{f(x)} - \frac{1}{L} \right| < \epsilon_0.$$

With a little algebra, this is

$$\left| \frac{f(x) - L}{f(x) \cdot L} \right| < \epsilon_0,$$

or

$$|f(x) - L| \cdot \left| \frac{1}{f(x)} \right| \cdot \left| \frac{1}{L} \right| < \epsilon_0.$$

Here's the plan. First, choose $\delta_1 > 0$ so that if $0 < |x - b| < \delta_1$ then $f(x) \geq L/2$. Then of course $1/f(x) \leq 2/L$ if $0 < |x - b| < \delta_1$. Now choose $\delta_2 > 0$ so that if $0 < |x - b| < \delta_2$ then $|f(x) - L| < \frac{\epsilon_0 \cdot L^2}{2}$ (what must be inserted into the definition of the limit of f to achieve this?). We now have two δ's; show the usual thing works, templates and all. (Careful: is $1/f$ ever undefined?)

6.2 Two δ's: The Sum Theorem

We return to the sum theorem: recall that we had arrived at

Pf. Let f_0 and g_0 be arbitrary functions with limits L and M at b, respectively.

Let $\epsilon_0 > 0$ be arbitrary.
Let $\delta_* =$ (**inspiration needed here!**).
. . .
Therefore we have found a δ_* to go with ϵ_0 such that
$\forall(x)(0 < |x - a| < \delta_* \Rightarrow |(f_0 + g_0)(x) - (L + M)| < \epsilon_0).$

So for ϵ_0, there does exist a $\delta_* > 0$ such that $\forall(x)(0 < |x - a| < \delta_* \Rightarrow |(f_0 + g_0)(x) - (L + M)| < \epsilon_0)$. Thus, since $\epsilon_0 > 0$ was arbitrary, $\forall(\epsilon > 0)\exists(\delta > 0)(\forall(x)(0 < |x - a| < \delta \Rightarrow |(f_0 + g_0)(x) - (L + M)| < \epsilon))$.
So $\lim_{x \to b} f_0(x) + g_0(x) = L + M$ as desired.
Since f_0 and g_0 were arbitrary functions, the result holds in general. ■

The hope was to use the ability of f and g to generate δ's to come up with a candidate for the δ for $f + g$. Here's the simple try.

Proof Discovery

Let $\epsilon_0 > 0$ be arbitrary. We need $\delta_* > 0$ so that $\forall(x)(0 < |x - b| < \delta_* \Rightarrow |(f_0 + g_0)(x) - (L + M)| < \epsilon_0)$. We know that $\forall(\epsilon > 0)\exists(\delta > 0)(\forall(x)(0 < |x - b| < \delta \Rightarrow |f(x) - L| < \epsilon))$. Also, $\forall(\epsilon > 0)\exists(\delta > 0)(\forall(x)(0 < |x - b| < \delta \Rightarrow |g(x) - M| < \epsilon))$. Simplest is to hand each of these ϵ_0, so let's. We get δ_1 and δ_2 such that

$$(6.6) \qquad \forall(x)(0 < |x - b| < \delta_1 \Rightarrow |f(x) - L| < \epsilon_0),$$

and
$$(6.7) \qquad \forall(x)(0 < |x - b| < \delta_2 \Rightarrow |g(x) - M| < \epsilon_0).$$

Trying $\delta_* = \min(\delta_1, \delta_2)$ is the standard way to combine two δ's.

We need $\forall(x)(0 < |x-b| < \delta_* \Rightarrow |(f+g)(x)-(L+M)| < \epsilon_0)$. So let x_0 be arbitrary such that $0 < |x_0 - b| < \delta_*$; we want $|(f+g)(x_0) - (L+M)| < \epsilon_0$. Since $\delta_* < \delta_1$, we have from $0 < |x_0 - b| < \delta_* < \delta_1$ that

$$(6.8) \qquad |f(x_0) - L| < \epsilon_0.$$

Similarly,
$$(6.9) \qquad |g(x_0) - M| < \epsilon_0.$$

Therefore,

$$(6.10) \quad \begin{aligned} |(f + g)(x_0) - (L + M)| &= |(f(x_0) - L) + (g(x_0) - M)| \\ &\leq |f(x_0) - L| + |g(x_0) - M| \\ &< \epsilon_0 + \epsilon_0 = 2\epsilon_0. \end{aligned}$$

This is, unfortunately, not what we want. It is close, but we need ϵ_0, not $2\epsilon_0$. To say something like "well, since ϵ_0 was *any* small number, $2\epsilon_0$ is *any* small number" is simply wiggling to avoid facing facts. This approach fails.

An analogy makes it clear why. If you nail together two things, each of which might be 1 inch too long or too short, couldn't you get something as much as 2 inches too long or too short? That is, each within ±1 inch is not enough to guarantee the result within ±1 inch, but only within ±2 inches. How would you force the final result within ±1?

6.19:

Returning to the proof, guaranteeing each of $|f(x_0) - L|$ and $|g(x_0) - M|$ less than ϵ_0 is not enough to guarantee $|(f + g)(x_0) - (L + M)| < \epsilon_0$. But a better guarantee on the individual terms succeeds. We got guarantees by handing ϵ_0 to each of the δ-producing limit definitions of f and g. In Section 6.1.2, we saw other choices. Make one, and show things work.

6.20:

Now write the proof carefully (quantifier templates, outline, candidate announcement for δ, and show it works, please).

Proof.

6.21:

6.2.1   Exercise

6.22: Consider again the proof above. Suppose we wanted not $|(f+g)(x_0)-(L+M)| < \epsilon_0$ but $|(f+g)(x_0) - (L+M)| < \epsilon_0/2$. (This device can be useful in more complicated proofs.) Can you arrange it by different inputs to the δ-producing machines for f and g? Do so. Can you cope with some number $K > 0$ and needing $|(f+g)(x_0) - (L+M)| < \epsilon_0 \cdot \frac{1}{K}$?

6.3   Two δ's (?): The Product Theorem

This is another (standard) "one new limit from two old limits" result; the technical details are trickier, but the template is the same.

Theorem 6.3.1 *Suppose f and g are functions satisfying $\lim_{x \to b} f(x) = L$ and $\lim_{x \to b} g(x) = M$. Then $f \cdot g$ satisfies $\lim_{x \to b}(f \cdot g)(x) = L \cdot M$.*

(Of course, $f \cdot g$ is defined by $(f \cdot g)(x) = f(x) \cdot g(x)$ for all x. Alternatively, the theorem is $\lim_{x \to b} f(x) \cdot g(x) = \lim_{x \to b} f(x) \cdot \lim_{x \to b} g(x)$, assuming both limits exist.) Start the scratchwork.

6.23:

The term $|f(x) \cdot g(x) - L \cdot M|$ in the scratchwork is rather difficult. We need the "smallness" of this term from the smallness of $|f(x) - L|$ and $|g(x) - M|$, but how isn't clear. The first of the technical tricks passed down from mathematician to mathematician is this: observe that

$$
\begin{aligned}
|f(x) \cdot g(x) - L \cdot M| &= |f(x) \cdot g(x) - f(x) \cdot M + f(x) \cdot M - L \cdot M| \\
&\leq |f(x) \cdot g(x) - f(x) \cdot M| + |f(x) \cdot M - L \cdot M| \\
&= |f(x)| \cdot |g(x) - M| + |M| \cdot |f(x) - L|.
\end{aligned}
$$

Can you force this last expression to be less than ϵ_0? One of the terms less than ϵ_0, at least?

6.24:

We use some ideas we have seen previously. First, we'd better make each of the two terms less than $\epsilon_0/2$ (cf. the sum theorem). Also, the second of the terms above is exactly as in Section 6.1.2 in the proof of the "constant multiplier" limit theorem (with c replaced by M). So we need to choose some δ_1 forcing $|f(x)-L| < \epsilon_0/(2|M|)$, but we can do that. Take a moment to let all these ideas solidify.

6.25:

The first term is harder. We want $|g(x) - M|$ so small that no matter what $f(x)$ multiplies it, the result is still less than $\epsilon_0/2$, but this isn't straightforward because there are many values of $f(x)$. We need the result in Exercises 6.14 and 6.15:

Lemma 6.3.2 *Suppose* $\lim_{x\to b} f(x) = L$. *Then there is a punctured neighborhood of b and a K so that $|f(x)| \le K$ for all x on the punctured neighborhood.*

Either do, or review, these Exercises from Section 6.1.3.[2]

6.26:

Everything is assembled. We need δ_1 for f so that f is bounded above by (some value) K on the punctured neighborhood of b given by δ_1. That done, we need δ_2 for g so $0 < |x-b| < \delta_2$ implies $|g(x) - M| < \epsilon_0/(2K)$.[3] We also get δ_3 for f so for $0 < |x-b| < \delta_3$ we have $|f(x) - L| < \epsilon_0/(2M)$. We now have three values of δ, and take δ_* to be the minimum, $\delta_* = \min(\delta_1, \delta_2, \delta_3)$.

Write up this proof in full detail.

6.27:

Insert the "$\forall f$" and "$\forall g$" templates if you forgot, and check what you wrote against the outline below.

> Pf. Let f_0 and g_0 be arbitrary functions with limits L and M at b, respectively.

---

[2]Leaving scratchwork for one proof to assemble needed pieces is distracting, but that's the way it usually goes.

[3]Note: δ_1 <u>must</u> come first, since it comes with K and we can't get δ_2 without K, because we get δ_2 by inserting $\epsilon_0/(2K)$ into the definition of limit of g.

> Let $\epsilon_0 > 0$ be arbitrary.
> Let $\delta_* =$ (**inspiration needed here!**).
> (**Your work here to show δ_* works.**)
> Therefore we have found a δ_* to go with ϵ_0 such that
> $\forall(x)(0 < |x - b| < \delta_* \Rightarrow |(f_0 \cdot g_0)(x) - (L \cdot M)| < \epsilon_0).$

So for ϵ_0, there does exist a $\delta_* > 0$ such that $\forall(x)(0 < |x - a| < \delta_* \Rightarrow |(f_0 \cdot g_0)(x) - (L + M)| < \epsilon_0)$. Thus, since $\epsilon_0 > 0$ was arbitrary, $\forall(\epsilon > 0)\exists(\delta > 0)(\forall(x)(0 < |x - b| < \delta \Rightarrow |(f_0 \cdot g_0)(x) - (L \cdot M)| < \epsilon))$.

So $\lim_{x \to b} f_0(x) \cdot g_0(x) = L \cdot M$ as desired.

Since f_0 and g_0 were arbitrary functions, the result holds in general. ∎

Take some time to look over (and if necessary improve) what you have written, since the payoff on that investment of time is considerable.

6.28:

Remarks

Congratulations; writing up this proof is a solid accomplishment (note we even snuck in a third δ). If you feel you aren't fully in control of things yet, or couldn't do this on your own, or that you are imitating without fully understanding, relax. It would be astonishing if you actually were in complete control. The techniques are new, the proof templates are new, and proving things at all is new. Remember that (almost) nobody fully gets these things the first time around, and this is a down payment on your future comprehension. Your understanding will grow with future encounters. Cheer up: what you would have written three weeks ago?

6.4 Two δ's: The Quotient Theorem

Theorem 6.4.1 *Suppose functions f and g satisfy $\lim_{x \to b} f(x) = L$ and $\lim_{x \to b} g(x) = M$, and also $M \neq 0$. Then $\frac{f}{g}$ satisfies $\lim_{x \to b}(\frac{f}{g})(x) = \frac{L}{M}$, or, alternatively,*

$$\lim_{x \to b} \frac{f(x)}{g(x)} = \frac{\lim_{x \to b} f(x)}{\lim_{x \to b} g(x)}.$$

This looks like an ordeal since quotients are generally worse than products. But Exercises 6.14 and following (the reciprocal theorem) save the day. Review these exercises or do them if you didn't before.

6.29:

We now get the quotient result almost immediately. Write the quotient $\frac{f(x)}{g(x)}$ as $f(x) \cdot \frac{1}{g(x)} = f(x) \cdot h(x)$ where $h(x) = \frac{1}{g(x)}$. By the reciprocal theorem, $\lim_{x \to b} h(x) = \frac{1}{M}$ if $M \neq 0$. By the product theorem, $\lim_{x \to b} f(x) \cdot h(x) = L \cdot \lim_{x \to b} h(x) = L \cdot \frac{1}{M} = \frac{L}{M}$ (still with $M \neq 0$). Done.

If you particularly like pain, you are free to try to prove the quotient theorem directly from the definition of limit without passing through the reciprocal theorem. (Pause.) Right. Good choice.

We now have the four very basic theorems: if f and g are functions and c a constant, we know the limits of $c \cdot f$, $f + g$, $f \cdot g$ and $\frac{f}{g}$ if we know about the limits of f and g separately. These will build a list of continuous functions efficiently, and we will get surprisingly far (for example, polynomials will be pretty painless).

6.4.1 Exercises

6.30: Return again to the sum theorem, but with *three* functions f, g, and h, with limits at b (respectively) L, M, and N. Prove, directly from the definition, that $\lim_{x \to b}(f(x) + g(x) + h(x)) = L + M + N$.

6.31: The limit operation is <u>linear</u>; this means that if f and g are any functions with limits L and M at b respectively, and c and d are any constants, then $\lim_{x \to b}(c \cdot f(x) + d \cdot g(x)) = c \cdot L + d \cdot M$. Prove this, by combining theorems and also directly from the definitions. Quantifiers in place, and cope with c or d zero, please.

6.32: Surprise: the limit of the product of three functions is the product of the three limits. Prove this, by combining theorems and also directly from the definitions by modifying the proof of the product theorem. [Hint: add and subtract more than one term.]

6.5 Two δ's: Composition

We work toward a limit result, but first review the definition itself.

6.5.1 Composition of Functions

Suppose that f and g are functions (defined for the moment on $\mathbf{R}$). For concreteness, let $f(x) = \sin x$ and $g(x) = x^2$ for all x. If we follow the usual conventions about parentheses, the symbol $f(g(4))$ is well defined: first take $g(4)$, i.e., 16, and insert it to get $f(16) = \sin 16$. Fine. Equally well defined is $f(g(3)) = f(9) = \sin 9$. In fact, for any x, $f(g(x)) = \sin x^2$. All is well.

This isn't wrong but is extremely "formula" based (exercises for this might include creating new formulas for $f(g(h(x)))$ given those for f, g, and h, or decomposing some complicated expression into a chain of formulas). We need more.

The generalization to two functions of a picture in Section 1.3 (the "domain-range" picture) helps.

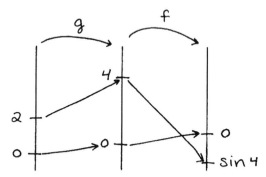

Some sample computations help, too. Realize that g comes first and is followed by f. For example, if $x = 2$ is in the domain of g, it is connected by a g arrow to 4 in the range of g. That 4 is in the domain of f, and so that 4 is connected by an f arrow to $\sin 4$ ($\approx -.757$, if you insist) in the range of f. For at least two other numerical values in the domain of g, draw the relevant g arrows and then the relevant f arrows. Also, for at least one value in the domain of f *not* in the range of g, draw the f arrow.

6.33:

You must keep track of "where you are." There is a 4 in the domain of g, another in the range of g and the domain of f, and yet another in the range of f. Only one of these, however, is naturally associated with the 2 in the domain of g. Similar care is needed with 0 in the domain of g. Try it.

6.34:

We must add to the diagrams above. For the functions above, surely $f(g(x)) = \sin x^2$ for all x, and we certainly ought to have a name for the function whose formula is $\sin x^2$. That name is $f \circ g$. Now there are f arrows, g arrows, and some new $f \circ g$ arrows. Below we show $(f \circ g)(2)$. You draw in the $f \circ g$ arrows for the values you followed previously.

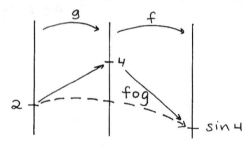

Compare the arrow indicating $f \circ g(1)$ to those for $g(1)$ and the f arrow from $g(1) = 1$.

Analogy I

One analogy suggested by the diagram is to riding subways or a bus system with transfers from one line to another. To get from point A to point B, you might ride the "g" line from A to C. You then hop on the f line, which goes from C to B, and there you are. Again, you must keep track of where you are: there is probably no point in getting on the f line at A, since *i)* the f line may not even leave from A, and *ii)* even if it does, it may not go from A to B.

Similarly, taking the g line once you get to C isn't right.[4] Note that the $f \circ g$ function is simply the result of the whole trip, getting from A to B on the subway (omitting the details of transfers).
End Analogy I

Analogy II

Another analogy is to a system of pipes (thin ones!) carrying water from place to place. A pipe from each point b of the domain of g carries water to somewhere in the range of g, say c; water entering at b into the g system comes out at c. There is also the f set of pipes: any point d in the domain of f connects to some point e in the range of f. Since pipes connect, if water flows from A to C in the g pipes, and flows from C to B in the f pipes, then water pumped in at A will come out at B; that flow corresponds to $f \circ g$.
End Analogy II

There is a new function, namely $f \circ g$, with arrows from the domain of g to the range of f. A definition is coming, but an example shows a potential problem: suppose we consider g above, with $g(x) = x^2$, but change f to be $f(x) = \frac{1}{x}$. If we try to consider $f \circ g(0)$, $g(0) = 0^2 = 0$, and $f(0)$ is undefined. We relaxed our rules about f and g being defined at all points, and we got into trouble. What about $g(x) = x^2 + 1$, and $f(x) = \frac{1}{x}$?

---

[4]You know that the g line does have an end at C, but it might not leave from C (end of the line?), and if it does, need g then go to B? No.

6.35:

So the problem isn't that f is not defined everywhere, but that g has in its range some point not in the domain of f. (Say this in terms of each of the analogies above.) Here is the definition that avoids the problem.

Definition 6.5.1 *Let f and g be functions such that the range of g is contained in the domain of f. Then we may define the function $f \circ g$ by $f \circ g(x) = f(g(x))$ for all x in the domain of f.*

The exercises below focus on understanding rather than computation.

6.5.2 Exercises

6.36: Suppose g is defined by $g(x) = \frac{1}{x}$ for all x not 0, and $f(x) = \frac{1}{x}$ for all x not 0. (Of course f and g are the same, but different names will help.) Draw diagrams and explore numerical values for $f \circ g$, and guess a "formula" for $f \circ g$. Is it consistent with what you get with manipulations at the formula level? What about $g \circ f$? Can you construct other g and f so $f \circ g(x) = x$ for all x?

6.37: Suppose $g(x) = x$ for all x, and f is any function whatsoever. What about $f \circ g$ and $g \circ f$? Careful: domains of definition!

6.38: Consider $g(x) = x^2$ for all x, and $f(x) = \sqrt{x}$ for all $x \geq 0$. What can you say about $f \circ g$ and $g \circ f$? Be careful!

6.39: Suppose we have three functions to hook together in sequence, say f, g, and h. Picture? Construct a good concrete example. What about $f \circ (g \circ h)$ vs. $(f \circ g) \circ h$?

6.5.3 Limits of Composite Functions

Suppose we have f and g so $f \circ g$ is defined. Suppose $\lim_{x \to b} g(x) = L$ and $\lim_{x \to L} f(x) = M$. Draw the relevant diagram (cf. Section 1.3).

6.40:

The limit of $f \circ g$ at b seems intuitively clear. "If x gets close to b, then $g(x) = y$ gets close to L, and if y gets close to L, f(y) gets close to M, so as x gets close to b we have that $f \circ g(x) = f(g(x))$ gets close to M." As usual, we illustrate a difficulty with an example. Let $g(x) = 5$ for all x, and let f be defined by

$$f(x) = \begin{cases} 3, & x \neq 5, \\ \text{undefined}, & x = 5. \end{cases}$$

To prove that $\lim_{x\to 2} f \circ g(x) = 3$ as hoped,[5] we need that for every $\epsilon > 0$ there exists $\delta > 0$ so for all x, $0 < |x-2| < \delta \Rightarrow |f \circ g(x) - 3| < \epsilon$. Take $\epsilon = .1$ and try $\delta = .1$. We hope that for all x, $0 < |x - 2| < \delta \Rightarrow |f \circ g(x) - 3| < \epsilon$. So let $x_0 = 2.01$, and note that indeed $0 < |x_0 - 2| < .1$. So we have to check $|f \circ g(x_0) - 3| < .1$. But $f \circ g(x_0) = f(g(x_0)) = f(g(2.01)) = f(5) = \ldots$ oops. Undefined. Check that no other value of δ could work better. While you are at it, show f is differently (but just as) bad if

$$f(x) = \begin{cases} 3, & x \neq 5, \\ 117, & x = 5. \end{cases}$$

6.41:

The problem is that values of g near b might land on L. And central to $\lim_{x\to L} f(x)$ is that we are not responsible for what f does at L; the "$0 < |x - L|$" portion of things ensures $x = L$ is never considered. Draw the picture of a g, and an f, for which these things might combine to give us trouble.

6.42:

So in coping with $f \circ g$, we must require more of f than usual.[6]

Theorem 6.5.2 *Suppose f and g are functions such that the composition $f \circ g$ is defined, $\lim_{x\to b} g(x) = L$, and f is continuous at L with $f(L) = M$. Then $\lim_{x\to b} f \circ g(x) = M$.*

Show that f continuous at L eliminates the problems above, where we merely assumed that f had a limit at L.

6.43:

We turn to the proof; remember that there are three functions, namely f, g, and $f \circ g$. We need, for some ϵ_0, a δ candidate for $f \circ g$. Aas usual, we use the ability of f and g to produce δ's. The diagram including f, g, and $f \circ g$ is useful.

Proof Discovery

Let $\epsilon_0 > 0$ be arbitrary to show that for every $\epsilon > 0$ there exists $\delta > 0$ such that for all x, $0 < |x - b| < \delta \Rightarrow |f \circ g(x) - M| < \epsilon$. Consider the

---

[5]What else, since obviously the limit of f at any point is 3?

[6]Another option (uncommon) is to ensure values of g are never equal to L, or at least never in some neighborhood of the point b.

diagram below with f, g, $f \circ g$, b, L, M, and the ϵ_0 interval around M shown. We ought to insert ϵ_0 into one of the δ-producing machines of f and g. (OK, $\epsilon_0/2$ if necessary, but simplicity first.) Which machine, f's or g's?

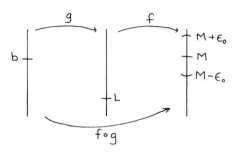

6.44:

If we are scrupulous about keeping track of "where we are," we ought to insert ϵ_0 into the definition of the limit of f, because the interval defined by ϵ_0 must be in the range of $f \circ g$, and so the set that is the range of g and domain of f is wrong. For every $\epsilon > 0$, there exists $\delta > 0$ so for all x, $0 < |x - L| < \delta \Rightarrow |f(x) - M| < \epsilon$. Insert ϵ_0 into this, and get δ_1 so for all x, $0 < |x - L| < \delta_1 \Rightarrow |f(x) - M| < \epsilon_0$. This leads to the diagram below with δ_1 shown.

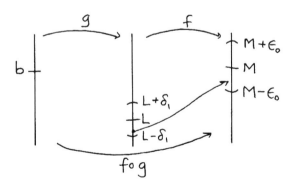

Note: we still lack δ around the point b, and have ignored g. Now what?

6.45:

The next step is new, but sensible. We ought to insert something into the definition of limit of g, probably a small number defining an interval in the range of g; we'll get a small number defining a (punctured) interval about b in the domain of g. But we *have* a small number defining an interval in the range of g, namely δ_1. For every $\gamma > 0$, there exists $\theta > 0$ such that for all x such that $0 < |x - b| < \theta$ we have $|g(x) - L| < \gamma$.[7] So inserting δ_1 in for γ, there is a $\theta_2 > 0$ so for all x such that $0 < |x - b| < \theta_2$ we have $|g(x) - L| < \delta_1$. We propose θ_2 as our candidate for the δ to go with ϵ_0 and $f \circ g$.

Take a moment to collect your thoughts. Then finish the picture, and complete the proof by showing δ works.

6.46:

6.5.4   Exercises

6.47: If we assume g is continuous at b (vs. just having a limit), what holds for $f \circ g$ at b? [Hint: What is $f \circ g(b)$?]

6.48: Prove the following straight from definitions. A picture helps.

Theorem 6.5.3 *Suppose f, g, and h are functions so that the composition $f \circ g \circ h$ is defined, such that $h(b) = c$ and h is continuous at b, $g(c) = d$, g is continuous at c, $f(d) = e$, and f is continuous at d. Then $f \circ g \circ h$ is continuous at b.*

6.6   Three δ's: The Squeeze Theorem

Don't panic: we will start with a two-δ version of the Squeeze Theorem and then improve it. Even before that, we motivate the theorem with some examples and an analogy.

Consider $f(x) = |x|$ and $g(x) = -|x|$ for all x. Graph these on the same set of axes.

6.49:

Now suppose h is a function always "between" f and g, which imprecise statement means that for all x, $g(x) \leq h(x) \leq f(x)$. (So, for example, $f(2) = 2$ and $g(2) = -2$, so $-2 \leq h(2) \leq 2$.) Note that equality of h with

---

[7]We changed the names so it is easier to insert δ_1 in for γ: this relieves psychological stress and changes nothing mathematically.

f and/or g is allowed. Add such an h to your graph, and don't be timid about making h wiggle around.

6.50:

Note $\lim_{x \to 0} f(x) = 0$ and $\lim_{x \to 0} g(x) = 0$. What's $\lim_{x \to 0} h(x)$?

6.51:

Yes, the values of f, g and h are all 0 at $x = 0$ (f and g by definition, and h because there is nowhere else legal). But that's irrelevant: this is a <u>limit</u> result, preserved if we redefine f, g, and h so that f and g are undefined at 0 and as before elsewhere, and h can do anything at all at 0. Redraw the diagram to fit, and take full advantage of h's freedom at $x = 0$ (now we require that for all x except possibly $x = 0$, $g(x) \le h(x) \le f(x)$). What about $\lim_{x \to 0} h(x)$ now?

6.52:

Clearly we can only say something about h because f and g "come together": consider $f_1(x) = |x| + 1$ and $g_1(x) = -|x| - 1$. Graph these, and find various h with various limits, or even none at all, still satisfying $g_1(x) \le h(x) \le f_1(x)$ for all x.

6.53:

Analogy
Here's another analogy that can stay private, but this is the football player theorem. Suppose you are walking between two very large, strong people, all of you marching in stride. You needn't be jammed up against each other, necessarily; they may be spread far apart with you in the middle, or far apart with you right next to either of them, or as you march along you can drift from near one to near the other, but you must remain between them. If they grow closer together or farther apart things can still go well for you, but if they go through a narrow door ... so do you.
End Analogy

This analogy clarifies why the example above can mislead you into a sort of "bow tie" theorem, which is much too limiting. Suppose that you and the two large strong people walk home in a wide field after, perhaps, a little too much root beer on a Saturday night. Perhaps you lurch from one side of the field to the other, either all three of you, or the two large people with you bouncing back and forth in between, so the progress toward the narrow door isn't like the absolute value example, where *all* values of f are larger than *all* values of g, but there is only a point-by-point $g(x) \leq h(x) \leq f(x)$.

Point by point is good enough. Consider $f_2(x) = |x| + x \sin x$ and $g_2(x) = -|x| + x \sin x$ for all x. Graph these (calculator!), and various h between them, and note that h is still "squeezed" to have limit 0 at $x = 0$.

6.54:

Modify once more to give a picture in which the three functions approach each other, and then retreat a little (rather than lurching in unison), finally pinching together at 0 as usual.

6.55:

Exercise

6.56: One more point: explain why $f(x) = .9|x| + \sin x$ and $g(x) = -.9|x| - \sin x$ aren't a suitable Squeeze Theorem pair.

Here's a careful statement of the Squeeze Theorem (first version).

Theorem 6.6.1 (Squeeze Theorem) *Let f, g, and h be functions defined for all x except possibly $x = b$, satisfying $g(x) \leq h(x) \leq f(x)$ for all x except $x = b$, and such that $\lim_{x \to b} f(x) = L$ and $\lim_{x \to b} g(x) = L$. Then $\lim_{x \to b} h(x) = L$ (that is, the limit exists, and is furthermore equal to L).*

Proof Discovery

As usual, to show that $\lim_{x \to b} h(x) = L$ we begin with $\epsilon_0 > 0$ arbitrary. We need a δ (for h, of course), and we expect that our candidate will come from the δ's arising from inserting something into the limit definitions of f and g. The straightforward thing is to insert ϵ_0 into them, so let's try that. Inserting into the limit definition of f at b, we get δ_1 so $\forall(x)(0 < |x - b| < \delta_1 \Rightarrow |f(x) - L| < \epsilon_0)$. Similarly, via g, we get δ_2 such that $\forall(x)(0 < |x - b| < \delta_2 \Rightarrow |g(x) - L| < \epsilon_0)$.

Try $\delta_* = \min(\delta_1, \delta_2)$ (the standard trick). Let x_0 be arbitrary such that $|x_0 - b| < \delta_*$. What must we show? Using what?

6.57:

It may be unclear how to combine the absolute value inequalities with the information that $g(x_0) \leq h(x_0) \leq f(x_0)$. The f inequality says $|f(x_0) - L| < \epsilon_0$, assuming that $|x_0 - b| < \delta_1$. Is $|x_0 - b| < \delta_1$? Why?

6.58:

We may expand $|f(x_0) - L| < \epsilon_0$ into $-\epsilon_0 < f(x_0) - L < \epsilon_0$. And since $h(x_0) \leq f(x_0)$, we have $h(x_0) - L \leq f(x_0) - L$. Judicious use of parts of these yields, for $h(x_0)$ vs. ϵ_0, ...

6.59:

Put this useful result on hold, and get another using g. Be sure to verify that $|x_0 - b| < \delta_2$! The pair of inequalities involving $h(x_0) - L$ can be assembled into the inequality we need involving $|h(x_0) - L|$.

6.60:

Note that the straightforward attempt in which we inserted ϵ_0 into the δ-producing machines for f and g (as opposed to some more complicated expression involving ϵ_0) worked fine.

Now write the proof, as opposed to its discovery. (Note in passing that an "$\epsilon_0/2$" or similar insertion wasn't needed.) Universal quantifiers on f and g, please, and check your argument against the form in Section 5.2.

6.61:

Pause for Breath
The writeup of such a proof is a fine achievement. Shortly we are going to complicate things a little, so take some time to enjoy what you've done.
End Pause

This section began with a *three δ* threat, but the above has only two. The third arises from an effort to improve the theorem. In terms of our analogy, we required that you have walked between the two football players for all past time and will for all future time (we require $g(x) \leq h(x) \leq f(x)$ for all x, x viewed as time t). How you want to spend your time is your business, but this is a needlessly strong hypothesis. If at some point you joined the large, strong people, then they went through the narrow door, then you stuck with them for a little longer, and finally you went your separate ways, clearly you went through the door too. So if $g(x) \leq h(x) \leq f(x)$ holds (only) for all x in some punctured neighborhood of b, that should be enough.

Graph, using $f(x) = |x|$ and $g(x) = -|x|$, and an h assumed between them only "near zero" to see how this looks.

6.62:

Enter the third δ: we'll assume that for some $\delta_3 > 0$, $g(x) \leq h(x) \leq f(x)$ for all x such that $0 < |x - b| < \delta_3$. To obtain $g(x_0) \leq h(x_0) \leq f(x_0)$ we now have to be sure that x_0 satisfies $0 < |x_0 - b| < \delta_3$, and all we assume is $0 < |x_0 - b| < \delta_*$. To get what we need we use the standard trick to ensure $\delta_* \leq \delta_3$, namely $\delta_* = \min(\delta_1, \delta_2, \delta_3)$.

Theorem 6.6.2 (Squeeze Theorem II) *Let f and g be functions each with limit L at b, and suppose that h is a function such that, on some punctured neighborhood N of b, $g(x) \leq h(x) \leq f(x)$ for all x in N. Then $\lim_{x \to b} h(x) = L$ (i.e., the limit exists, and is furthermore equal to L).*

Write up the proof using the techniques above.

6.63:

The replacement of "global" hypotheses by "local" ones exploits the fact that "limit" is a local notion. We've used localization previously in Exercises 6.14 and following (when showing, for example, that if $\lim_{x \to b} f(x) = L > 0$ then locally f is nonzero). The exercises below revisit other previous results to improve them in this way.

6.6.1 Exercises

6.64: Consider the result of Proposition 6.1.1. We assumed that the functions were defined on all of **R**, and even equal on all of **R**, but that was unnecessary. Here is a statement of a better result.

Proposition 6.6.3 *Let f and g be functions so that on some punctured neighborhood N of b, the equality $f(x) = g(x)$ holds for all x in N. Then if $\lim_{x \to b} f(x) = L$ we have $\lim_{x \to b} g(x) = L$.*

Prove this proposition by modifying the proof we gave for the original and including another δ to localize to those x on which we can count on the behavior of f and g. It will probably help to draw a picture of an f and g that coincide on some neighborhood of a point, but not on the *whole* real line excluding that point.

6.65: Improve the result of Exercise 6.13 by removing the hypothesis that g is bounded above and below everywhere and replacing it with a suitable localization to a punctured neighborhood of b. Prove the result.

6.7   Change of Limiting Variable

Consider $\lim_{x \to 5} x^2$. Construct a table of input and output values of the function near, and on both sides of, 5 (yes, it is foolishly simple, but we need it for what is coming next). Also, consider $\lim_{z \to 0}(5+z)^2$, and construct a table of values for inputs near, and on both sides of, 0.

6.66:

No surprise: each of the functions has limit 25, the first as x gets close to 5, the second as z gets close to 0. But while lots of functions have limit 25 at different points, that the limit of x^2 at $x = 5$ is 25 is intimately connected to the other limit. For example, if $x = 5.1$ the value of the first function is 26.01, and if the value of z is .1 the value of the second function is also 26.01. Find other such "coincidences."

6.67:

There's nothing deep and mysterious here. Let $f(x) = x^2$. We've noticed various numerical examples of the following fact: if $z = x - 5$, then $f(x) = f(5+z)$. The graph below indicates the geometric meaning of this algebraic fact. The limit is either "what happens to f when inputs get close to 5" or as "what happens to f when the difference between our input and 5 gets close to 0."

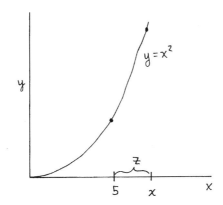

There is some new notation. This difference between input and limit point ("z" above) is customarily written Δx. The first symbol is the capital Greek letter Δ, read "delta," and is used in mathematics to denote a (usually small) change in a value, so Δx is a small change in the value x. With this language we may state the proposition.

Proposition 6.7.1 *Let f be a function and b a real number. Then*

$$\lim_{x \to b} f(x) = \lim_{\Delta x \to 0} f(b + \Delta x),$$

meaning that if either limit exists, both do, and their values are equal.

We leave the proof of this proposition to you; it is a rather simple example of the sort of "single δ" proof which we considered in Section 6.1.[8]

6.7.1   Exercise

6.68: Prove Proposition 6.7.1.

---

[8]We could have considered it much earlier, but the point of this technical result won't be seen until we consider the limits of trigonometric functions, and later when we discuss the derivative.

7

Which Functions are Continuous?

We return to the question of which functions are continuous (at a point, and on various sets — recall from Sections 2.1, 2.3, and 2.3.2 the multiple meanings of "continuous"). Some results are easy; with a new proof form, we'll get continuity of polynomials.

7.1 Preview: Direct from the Limit Theorems

Our limit theorems for "sums," "products," and so on, give some continuity results easily. Here's a sample, with the rest in the exercises.

Theorem 7.1.1 *Let f and g be functions continuous at a real number a. Then $f + g$ is continuous at a.*

Proof. Suppose f, g, and a are as in the hypothesis.[1] Then

$$
\begin{aligned}
(f + g)(a) &= f(a) + g(a) \\
&= \lim_{x \to a} f(x) + \lim_{x \to a} g(x) \\
&= \lim_{x \to a} (f(x) + g(x)) \\
&= \lim_{x \to a} (f + g)(x),
\end{aligned}
$$

---

[1] Warning: the universal quantifier templates are in use for f, g, and a. Explicitly? No. Implicitly? Yes.

where the second equality is from continuity at a for f and g, the third equality is from the result on the limit of a sum from Section 6.2, and all other equalities are from the definition of the sum of two functions.

7.1.1   Exercises

7.1: Prove that the product of continuous functions is continuous.

7.2: Repeat for quotients, excluding one case you should state precisely.

7.3: Prove that a constant multiple of a continuous function is continuous.

7.4: Prove that if f and g are continuous and c and d are any constants, then $cf + dg$ is continuous. This indeed combines earlier results, but records that continuity behaves well under "linear combinations."

7.5: Prove that if f and g are functions so that $f \circ g$ is defined, if g is continuous at a, and if f is continuous at $g(a)$, then $f \circ g$ is continuous at a.

7.2   Small Integer Powers of x

The ϵ-δ arguments for the following are easy, because the functions are simple. Give the proofs.

Proposition 7.2.1 *Let c be any constant; then the function f defined by $f(x) = c$ for all x is continuous on* $\mathbf{R}$*. Further, the function g defined by $g(x) = x$ for all x is continuous on* $\mathbf{R}$*.*

7.6:

Remark that $f(x) = x^0$ for all x is not quite the constant function 1 (0^0 is the only problem); ignoring this, we proved x^0 and x^1 are continuous functions. Answer the obvious next question.

7.7:

We've done x^2 (at least at a few points). An ϵ-δ proof in general would build technical strength, but work. Look ahead, though: ϵ-δ proofs for x^3 and x^4? How about for x^{1937}?

We worked pretty hard for "new limits from old" theorems in Chapter 6, and just got continuity versions. One payoff is the following quick and easy proof.

Proof.   Let f be defined by $f(x) = x$ for all x, and $g(x) = x^2$. Note $g = f \cdot f$, and we just showed f is continuous at each point of $\mathbf{R}$. To show g is continuous, let b in $\mathbf{R}$ be arbitrary. Then

(7.1)
$$
\begin{aligned}
g(b) &= b^2 \\
&= b \cdot b \\
&= f(b) \cdot f(b) \\
&= \lim_{x \to b} f(x) \cdot \lim_{x \to b} f(x) \\
&= \lim_{x \to b} f(x) \cdot f(x) \\
&= \lim_{x \to b} g(x).
\end{aligned}
$$

Since b was arbitrary, g is continuous at each point of $\mathbf{R}$.

Justify each step, please. Then, prove x^3 is continuous using continuity of x^2. You'll need a universal template for "all points in the domain."

7.8:

These are nice results, and surprisingly painless. Lesson: theorems of the very general variety we proved in Chapter 6 were hard (e.g., had lots of quantifiers) exactly because they are strong theorems. That work avoided ϵ-δ for x^2 at each point of its domain, then all over again for x^3, and so on. More rewards are coming.

Question: is x^{1937} continuous at each real number, and if so, why?

7.9:

Obviously yes, but the "why" part is harder. Intuitively, surely x^3 is continuous based on the above proof and the continuity of x^2 and x, and then surely x^4 is continuous similarly, and keep going until we have worked up to x^{1937}. In the next section, we'll formalize this intuition that one can keep going into a proof method called mathematical induction. This will yield continuity of all the powers x^n, and a lot more.

7.2.1   Exercises

7.10: Prove that $f(x) = \frac{1}{x}$ ($x \neq 0$) is continuous (i.e., continuous on its domain). Then prove, in two ways, that $g(x) = \frac{1}{x^2}$ ($x \neq 0$) is continuous.

7.11: Assume temporarily that each power x^n, $n = 1, 2, \ldots$ is continuous. Prove that each monomial (cx^n for some c) is continuous. Prove that a sum of two monomials is continuous; prove that the sum of three monomials is continuous, based on the length two sum. To get the result for any length sum, again we need to "keep going."

7.12: Prove in two essentially different ways that $h(x) = (x+1)^2$ is continuous. Repeat for $i(x) = \dfrac{1}{(x+1)}$ $(x \neq -1)$.

7.13: Prove appropriate continuity statements for

$$f(x) = \frac{x^2 + 2}{x^2 - 1}.$$

7.3   Mathematical Induction

In Section 5.1.5 we discussed a template for the proof of things quantified by a universal quantifier. That template is general, hence widely applicable. The price is that it is unspecialized. Mathematical induction, another way to prove universally quantified statements, is limited to proving statements quantified over "all positive integers." That specialization allows use of the fact that, in the natural ordering of the positive integers, there is a first element, and every other element has an element immediately prior to it. (This isn't true for, say, the set of all functions.) This special structure appears in the intuition that if you can prove something for 1, and then prove something for 2 using the fact for 1, and for 3 using the fact for 2, and so on, it ought to be true for all integers, since we can "keep going."

Theorem 7.3.1 (Induction Theorem) *Let $P(n)$ be a condition with n (a positive integer variable) the only free variable.[2] To prove $\forall n \geq 1(P(n))$ it is enough to prove*

  1. $P(1)$, (the "$n = 1$ step") and

  2. $\forall n \geq 1(P(n) \Rightarrow P(n+1))$ (the "induction step").

A sample "statement about n" is "the function f_n defined by $f_n(x) = x^n$ is continuous." So our theorem "for every positive integer n, the function f_n defined by $f_n(x) = x^n$ is continuous" is suitable for induction.

We detour from continuity to justify this theorem, and then return to the use of the Induction Theorem.

---

[2]Don't panic. This technicality is solely to make the mathematical logicians happy. Read instead, "Let $P(n)$ be a statement about n." This means that inserting any value in for n (like, 5) we get a statement unambiguously true or false.

7.3.1   Justification of the Induction Theorem

The following might be proved in a course in mathematical logic or set theory, but is beyond this text, however reasonable it seems. **Fact**: Any set of positive integers with at least one element contains a least element.

Assume this in what follows, and write **N** for the set of positive integers. Suppose then that we have some statement $P(n)$ to prove true for all n in **N**. Suppose we follow the Induction Theorem, namely, prove that $P(1)$ is true and that $\forall n \geq 1(P(n) \Rightarrow P(n+1))$. Why then must $P(n)$ be true for all n? Embark on a proof by contradiction, so assume that there is some n_0 in **N** so that $P(n_0)$ is false.

Consider the set S of all positive integers m such that $P(m)$ is false; it is not empty because of n_0. Therefore there is a least element in S, which we call n_*; n_* is the least n such that $P(n)$ is false.

First, n_* cannot be 1, because we have assumed that we have proved $P(1)$, and $P(1)$ can't be both true and false. Since n_* is not 1, $n_* - 1$ is a positive integer. Further, since $n_* - 1$ is less than n_*, $P(n_* - 1)$ must be true (since n_* is the *least* n for which $P(n)$ is false). But we have proved that for all n, if $P(n)$ is true then $P(n+1)$ is true. Applying this with n set to $n_* - 1$, we get that if $P(n_* - 1)$ is true then $P(n_* - 1 + 1)$, i.e., $P(n_*)$, is true as well. So we have found that $P(n_* - 1)$ is true; also, $P(n_* - 1)$ is true implies $P(n_*)$ is true.

Clearly, then, $P(n_*)$ is true, contradicting $P(n_*)$ false. So our assumption was false: there aren't n in **N** such that $P(n)$ is false, so $P(n)$ is true for all n in **N**, as desired.

This justification helps some people, but realize that correct use of Mathematical Induction doesn't rest on perfect comprehension of its proof. The use comes next, with some terrific results.

7.3.2   Use of the Induction Theorem

Here's a use of the Induction Theorem utterly irrelevant to what we are doing, but simple and illustrative because it is numerical in nature.[3] Consider the following proposition:

Proposition 7.3.2 *For all positive integers n, $1 + 2 + \ldots + n = \dfrac{n(n+1)}{2}$.*

(There's a little notational slack: if n is 1, the left-hand side means "1", not "$1 + 2 + 1$." We are after the sum of the first n positive integers.)

---

[3]It is also an induction standard, partly because this formula is involved in (some versions of) a story. See the entry on Gauss in [1], a famous source of tales about mathematicians (although sometimes a good story displaces accuracy, and unfortunately it was written before the realization that both genders could do mathematics).

Since universally quantified, with the set of allowed values $\mathbf{N}$, the Induction Theorem (or possibly the usual '$\forall$' template) is suitable. Note also that '$P(n)$' is $1 + 2 + \ldots + n = \dfrac{n(n+1)}{2}$. We are set up for a proof by induction, so here it is.

Proof.

1. We must first prove $P(1)$, i.e., that $1 = \dfrac{1(1+1)}{2}$, clearly OK.

2. Next is: for all $n \geq 1$, $P(n) \Rightarrow P(n+1)$. We will use the usual template for the proof of universally quantified things. (If confused, just press on.) So let $n_0 \geq 1$ be arbitrary: we need $P(n_0) \Rightarrow P(n_0 + 1)$. So assume $P(n_0)$, i.e., that

$$(7.2) \qquad 1 + 2 + \ldots + n_0 = \frac{n_0(n_0 + 1)}{2}.$$

We must show (probably using this assumption) that $P(n_0 + 1)$ holds, i.e.,

$$(7.3) \qquad 1 + 2 + \ldots + (n_0 + 1) = \frac{(n_0 + 1)((n_0 + 1) + 1)}{2}.$$

It does, since

$$
\begin{aligned}
1 + 2 + \ldots + (n_0 + 1) &= 1 + 2 + \ldots + n_0 + (n_0 + 1) \\
&= \frac{n_0(n_0 + 1)}{2} + (n_0 + 1) \\
&= \frac{n_0(n_0 + 1)}{2} + \frac{2(n_0 + 1)}{2} \\
&= \frac{(n_0 + 1)((n_0 + 1) + 1)}{2},
\end{aligned}
$$

where the first equality just makes explicit that n_0 is the second-to-last term in the sum up to $n_0 + 1$, the second equality is justified by using our hypothesis in equation (7.2) and substituting, and the rest is just algebra. Since n_0 was arbitrary, we're done with the induction step, hence done.

Let's turn to the most important of your (many?) questions: "If the whole idea of induction is to give an alternative to using the universal template, why do we use the universal template in the middle of our induction proof?" Alternatively, "isn't the thing we do in part 2 just the universal template we would have had to do anyway?"

Quite Crucial

Terrific questions. For the standard '$\forall$' template, we let n_0 be arbitrary and try to prove $P(n_0)$. As *part* of our induction form, we let n_0 be arbitrary and try to prove $P(n_0) \Rightarrow P(n_0 + 1)$. Assuming n_0 arbitrary is the same, but what we are proving is not. In one case, we must prove $P(n_0)$ from scratch, in the other we prove $P(n_0) \Rightarrow P(n_0 + 1)$, surely different.

The latter isn't always easier, but in the usual universal quantifier template we must prove $P(n_0)$ with no obvious tools to work with, while in the similar-seeming portion of the induction form we prove $P(n_0 + 1)$ *allowed to use the assumption $P(n_0)$*. In the proof above we *did* use the assumption $P(n_0) : 1 + 2 + \ldots + n_0 = \dfrac{n_0(n_0 + 1)}{2}$. Similarly, in proving x^2 continuous we used x continuous.

End Crucial

The exercises give worthwhile pattern practice, not vital results.

7.3.3 Exercises

7.14: Prove that for each n in **N**, $1^2 + 2^2 + \ldots + n^2 = \dfrac{n(n + 1)(2n + 1)}{6}$.

7.15: Prove that for all n in **N**, $1^3 + 2^3 + \ldots + n^3 = \left(\dfrac{n(n + 1)}{2}\right)^2$.

7.16: Prove for each n in **N** that $1 + 3 + 5 + \ldots + (2n - 1) = n^2$.

7.4 Better Use of the Induction Theorem: Polynomials Plus

Away with numerical sums, and on to powers of x.

Theorem 7.4.1 *For any n a positive integer, the function f_n defined by $f(x) = x^n$ is continuous at each real number, and hence continuous on* **R**.

Try the proof. Suggestion: write it as "at each real number b, and for each n in **N**, the function f_n defined by $f(x) = x^n$ is continuous at b." The '$\forall b$' template removes b problems.

7.17:

The $n = 1$ step was done in Proposition 7.2.1. Further, the "induction step" is indicated by the proof (in Section 7.2) that x^2 is continuous at an arbitrary point. If necessary, try again. Thereafter, a result follows almost for free.

7.18:

Corollary 7.4.2 *Let c be any constant and n be an arbitrary positive integer. Then the function f defined by $f(x) = cx^n$ is continuous on* **R**.

(This result is sometimes put in words by saying that each "monomial" is continuous (where a "monomial" is a function of the form cx^n).)

Proof.

7.19:

Pause for Breath

Pause before we tackle polynomials: the problem is moving from a single term to (potentially very long) sums of terms. Take a break first.

End Pause

Start by asking, "what is a polynomial?" One definition is "a sum of monomials," but more precision helps. A polynomial of degree n is a sum of monomials whose powers of x are all less than or equal to n and such that each such power occurs exactly once as a power in the sum (thus a sum of exactly $n+1$ monomials). Note also that some or all of the coefficients might be zero, so, for example, $4x^3 + 1$ shall be written $4x^3 + 0x^2 + 0x^1 + 1$. The fussiness will make sense soon.

Let's start with small n to see what we are getting ourselves into.

Proposition 7.4.3 *At any point b in $\mathbf{R}$, any polynomial of degree 1 is continuous.*

Proof.

7.20:

Good. Let's move on to a polynomial of degree two and see what happens.

Proposition 7.4.4 *At any point b of $\mathbf{R}$, any polynomial of degree 2 is continuous.*

Proof Discovery

A polynomial of degree 2 is $c_0 + c_1 x + c_2 x^2$. How can we show this is continuous at the arbitrary b we are surely going to pick? One approach is to see the sum of three continuous functions (the monomial terms), and look for a "sum of three functions" theorem. In Exercise 6.30 we did prove a "limit of the sum of three functions" result; a length-three sum continuity result could have followed (but didn't). This does work. But what is troublesome about the approach?

7.21:

Ready for a polynomial of length 23? Should we look forward to proving the limit of a sum of, say, length 23 is the sum of the 23 limits, using, probably, 23 values of δ? Yuk (a technical term). Worse, sums of *any* length will never be done this way. Crucial is that we want some general theorem about sums of all possible finite lengths n (like, "If each term in the sum is continuous so is the sum as a whole."). The "all possible finite lengths n" indicates the approach.

7.22:

Moral: proving the continuity result for the sum of length three ought to contain within it the seeds of the proof for all sum lengths, just as our proof that x^2 was continuous indicated the "all x^n" proof.

Note that our target has shifted from the polynomial of degree 2 (implicitly, all polynomials) to sums of *any* continuous functions where n is the "any length of sum" that stands in our way. Here is the new target precisely; prepare the Induction Theorem.

Theorem 7.4.5 *Let a be an arbitrary real number, n be an arbitrary positive integer, and let $f_1, \ldots, f_n$ be a collection of functions each of which is continuous at a. Then the function f defined by $f(x) = f_1(x) + \ldots f_n(x)$ for all x is continuous at a.*

Grasp that this is the new goal, and that it gives continuity of polynomials.

7.23:

To the proof; the $n = 1$ (a sum of length 1) is easy, right? The function f is just f_1 (only one term), and f_1 is assumed continuous at a, done. Now we'll assume that a sum of length n is continuous and try to prove that a sum of length $n + 1$ is continuous. We therefore face $f_1 + \ldots + f_{n+1}$ and have somehow to use the continuity of $f_1 + \ldots + f_n$. What is the key idea (look at the sums of powers-of-integers proofs)?

7.24:

We should write the sum $f_1 + \ldots f_{n+1}$ as $f_1 + \ldots + f_n + f_{n+1} = (f_1 + \ldots + f_n) + f_{n+1}$. This is a sum of two functions (fits a theorem!), one $f_1 + \ldots + f_n$, the other f_{n+1}. Is each continuous? Why? Specify the sources of your beliefs.

7.25:

Write up the proof; don't peek at the Hint too soon, because it contains a complete proof and further discussion to read afterwards.

7.26:

(Make sure you read the Hint even after you are done, since the discussion there is important.)

Well, that was a lot of work. But we get that polynomials are continuous (i.e., continuous at each point of the domain, which is $\mathbf{R}$ in this case) since each polynomial is a sum (of some length) of monomials, and monomials are continuous.

Corollary 7.4.6 *Any polynomial is continuous at all real numbers.*

7.4.1   Exercises

7.27: Suppose f_1, ..., f_n, ... are functions each known to be continuous at 2. Prove from scratch that any sum $\sum_{i=1}^{i=N} f_i$ is continuous at 2.

7.28: Prove a product (any number of terms) of continuous functions is continuous.

7.29: It is possible (if foolish) to get the continuity of polynomials without a general sum theorem; do the induction, based on the idea that a polynomial of degree $n+1$ is the sum of a monomial (continuous) and a polynomial of degree n that can be assumed continuous via the induction hypothesis.

7.30: As another alternative, prove by induction a result about the limit of a sum of arbitrary length.

7.4.2   Rational Functions

Trick question: how will we use induction to prove that rational functions (quotients of polynomials) are continuous where they ought to be?

7.31:

It is easy enough to get past this trick if you're clear about what induction is good for: proving something for each value of n, where n ranges over $\mathbf{N}$. And for polynomials, this was perfect because a polynomial is a sum of some length n in $\mathbf{N}$. But a rational function is a *single* quotient of polynomials, not 1 or 2 or n quotients of polynomials, so induction is not the right tool. We will use our rule about limits or continuity of quotients (see Section 6.4 or Exercise 7.2) and our recent polynomial results.

Try this, then: prove every rational function is continuous on $\mathbf{R}$.

7.32:

We hope you rejected this foolishness, but care is needed. "Continuous on $\mathbf{R}$" means "continuous at each point of $\mathbf{R}$," in particular "defined at each point of $\mathbf{R}$," a condition not guaranteed for rational functions. So the statement above is doomed.

All right, two trick questions is enough. Prove each rational function is continuous at each point of its domain. (Equivalently, using Definition 2.3.5, we are proving each rational function is continuous.)

Theorem 7.4.7 *A rational function is continuous.*

7.33:

7.4.3   Exercises

7.34: Prove that if $r_1, \ldots, r_n, \ldots$ are rational functions, any sum $\sum_{i=1}^{N} r_i$ is continuous. There are many approaches; try several.

7.35: Prove that if n is any integer, then f defined by $f(x) = x^n$ is continuous.

7.36: Assume temporarily that the sine function is continuous at each b in $\mathbf{R}$. Prove that for any n in $\mathbf{N}$, f defined by $f(x) = \sin^n x$ is continuous at each b in $\mathbf{R}$. Prove that any polynomial in the sine (that is, a sum of $c_j \sin^j x$) is continuous.

7.37: Assume temporarily that "e^x" is continuous on $\mathbf{R}$. Prove that if p is any polynomial, f defined by $f(x) = e^{p(x)}$ is continuous on $\mathbf{R}$.

7.5   Trigonometric Functions (Trouble)

[General Warning: we omit a complete presentation so as not to be drowned in technical details; the remainder of this chapter may be skipped without harm to your understanding of subsequent ones.]

What technical details? A graph of the sine function certainly seems continuous at every point.[4]

7.38:

---

[4]We take for granted the sine and other trigonometric functions as functions on $\mathbf{R}$, and use standard facts about them (e.g., trigonometric identities).

Further, via a trick using a standard trigonometric identity, continuity of the sine and cosine functions at 0 gives continuity everywhere. Justify each of the equality signs (Section 6.7 gives a justification of one).

$$
\begin{aligned}
(7.4) \quad \lim_{x \to a} \sin x &= \lim_{\Delta x \to 0} \sin(a + \Delta x) \\
&= \lim_{\Delta x \to 0} \sin(a) \cdot \cos(\Delta x) + \cos(a) \cdot \sin(\Delta x) \\
&= \lim_{\Delta x \to 0} \sin(a) \cdot \cos(\Delta x) + \lim_{\Delta x \to 0} \cos(a) \cdot \sin(\Delta x) \\
&= \sin(a) \cdot \lim_{\Delta x \to 0} \cos(\Delta x) + \cos(a) \cdot \lim_{\Delta x \to 0} \sin(\Delta x).
\end{aligned}
$$

7.39:

If the sine were continuous at 0, we would have $\lim_{\Delta x \to 0} \sin(\Delta x) = \sin 0 = 0$, and if the cosine were continuous at 0, we would have as well that $\lim_{\Delta x \to 0} \cos(\Delta x) = \cos 0 = 1$. Assuming these for the moment, we have

$$
\sin(a) \cdot \lim_{\Delta x \to 0} \cos(\Delta x) + \cos(a) \cdot \lim_{\Delta x \to 0} \sin(\Delta x) = \sin(a) \cdot 1 + \cos(a) \cdot 0 = \sin(a).
$$

Combining this with (7.4), we get exactly $\lim_{x \to a} \sin x = \sin(a)$, i.e., continuity of the sin at a.

Amazing: continuity of the sine and cosine at 0 yields the continuity of the sine at *any* point. Similarly one gets continuity of the cosine, and from the two of these the continuity of the other trigonometric functions. Pretty remarkable. Because of this, though, the continuity of the sine and cosine at 0 is more important than one would at first suspect. Since the sine and cosine functions are periodic, it seems perfectly reasonable that if the sine (say) is continuous at $\pi/2$ then it ought also to be continuous at $2\pi + \pi/2$. That is, continuity of the sine over a complete period (such as $[0, 2\pi]$) ought to guarantee continuity on any other period, and the same thing ought to work for the cosine. But to have continuity everywhere boil down to continuity at zero is surely using further structure, deeper than just the fact that these functions are periodic.

7.5.1 Continuity of the Sine at 0

To give the usual argument for the continuity of the sine at 0, we will use the Squeeze Theorem and some geometry. Recall the radian measure of an angle θ in standard position is the arc length of the portion of the unit circle cut off by θ, at least for angles θ in the interval $(0, \pi/2)$. Further, $\sin \theta$ is just the y coordinate of the relevant point on the circle. Below is the diagram capturing all this standard geometry.

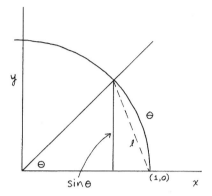

On geometrical grounds, θ is greater than $\sin\theta$. (Introduce ℓ, the relevant chord on the circle.) Also, clearly, $\sin\theta > 0$. We have therefore shown that

$$0 < \sin\theta < \theta, \ \ 0 < \theta < \pi/2.$$

Via the Squeeze Theorem (yes, a one-sided version — don't quibble) we get that the right hand limit of the sine at 0 is 0, since $\lim_{\theta\to 0+} 0 \le \lim_{\theta\to 0+}\sin\theta \le \lim_{\theta\to 0+}\theta$ and each of the squeezing functions has limit 0 at $\theta = 0$.

The limit from the left may be done using $\sin(-\theta) = -\sin\theta$. Then, since both the left- and right-hand limits of the sine at 0 are 0, we have finally that $\lim_{\theta\to 0}\sin\theta = 0 = \sin 0$. This is exactly continuity of the sine at 0.

7.5.2   What's the Problem?

The proof above is, to me, utterly convincing. In many ages of mathematics it would have been regarded as a perfectly good proof. But expectations for proofs changed in the 1800s, when it became required that a proof flow from a certain set of axioms (often based on properties of sets) in an orderly way. The appeals above to geometry, undefined "arc length" and so on fell into disfavor. The above proof isn't wrong, but there's a lot not actually justified. We might feel more comfortable with a proof based on calculus without all this geometry stuff. And that's why this proof can be perfectly convincing and still raise an eyebrow.

What's the solution? One approach is to define the sine function in a way not geometrically based. An infinite series definition (see Section 10.2) is

$$\sin x = x - \frac{x^3}{3!} + \frac{x^5}{5!} - \frac{x^7}{7!} + \cdots \ .$$

There are a few minor problems. Do these infinite sums even mean something? Is it obvious that $\sin^2 x + \cos^2 x = 1$ for all x? Among other things, our basic goal requires continuity (surprisingly relatively easy).

We'll continue based on this geometrical "proof" of the continuity of the sine at 0; that approach can be justified, although we won't justify it. If all the discussion above makes you nervous, you are free to behave as if we had assumed as an axiom that the sine function is continuous at 0.

7.5.3 A Lemma: Continuity of the Square Root

To prove continuity of the cosine at 0 we need a lemma. We prove as little as we can get away with, but it is fairly technical, and if you are content to accept its conclusion you may skip this section.

Lemma 7.5.1 *Let f be the function defined by $f(x) = \sqrt{x}$. Then f is continuous at each point of $(0, \infty)$.*

Proof Discovery

In fact, all we will prove is continuity at $b = 1$ (all we need for the cosine). Clearly the square root function is defined at 1 and in some neighborhood of 1, so we need $\lim_{x \to 1} \sqrt{x} = \sqrt{1}$. Let $\epsilon > 0$. We seek $\delta > 0$ so that if $|x-1| < \delta$ then $|\sqrt{x} - \sqrt{1}| < \epsilon$. It helps to notice that $|x - 1| = |\sqrt{x} - \sqrt{1}| \cdot |\sqrt{x} + \sqrt{1}|$.

Here's some scratchwork. If we have $|x - 1| < \delta$, we would have

$$|\sqrt{x} - \sqrt{1}| \cdot |\sqrt{x} + \sqrt{1}| < \delta,$$

so we would have

$$|\sqrt{x} - \sqrt{1}| < \frac{\delta}{|\sqrt{x} + \sqrt{1}|}.$$

This would be fine if, somehow, we had arranged to have

$$\frac{\delta}{|\sqrt{x} + \sqrt{1}|} \leq \epsilon.$$

So we'd have the naive choice of δ to be $|\sqrt{x} + \sqrt{1}| \cdot \epsilon$; checking, if

$$|x - 1| < |\sqrt{x} + \sqrt{1}| \cdot \epsilon,$$

then

$$
\begin{aligned}
|\sqrt{x} - \sqrt{1}| &= \frac{|\sqrt{x} - \sqrt{1}| \cdot |\sqrt{x} + \sqrt{1}|}{|\sqrt{x} + \sqrt{1}|} \\
&= \frac{|x - 1|}{|\sqrt{x} + \sqrt{1}|} \\
&< \delta \cdot \frac{1}{|\sqrt{x} + \sqrt{1}|} \\
&< |\sqrt{x} + \sqrt{1}| \cdot \epsilon \cdot \frac{1}{|\sqrt{x} + \sqrt{1}|} \\
&= \epsilon.
\end{aligned}
$$

This is, of course, a completely nonsensical and illegal choice of δ. Why?

7.40:

But there's a standard technique for fixing the presence of the "x" by eliminating it; choose a δ smaller than any possible $|\sqrt{x} + \sqrt{1}| \cdot \epsilon$. But since $\sqrt{1} = 1 < \sqrt{1} + \sqrt{x}$ for any x, $\delta = \epsilon$ is such a δ. To keep δ so small that none of the x in $(1 - \delta, 1 + \delta)$ could be negative (i.e., bad for the square root), we also need $\delta \leq 1$. So we will take $\delta = \min(1, \epsilon)$.

Write up the proof elegantly; explicit quantifiers, please.

7.41:

7.5.4 Exercises

7.42: Show that the square root function is right continuous at 0. Tackle this from scratch; imitating the proof above is not efficient.

7.43: (Challenge problem) Complete the proof that $\sqrt{x}$ is continuous on $(0, \infty)$. The proof above works, with some 1's changed to b's.

7.5.5 Continuity of the Cosine at Zero

Recall that this is the last piece needed to get continuity of all the trig functions; amazingly, continuity of the sine and cosine at zero would establish their continuity everywhere. We got continuity of the sine at 0 via some geometrical trickery, and continuity of the square root at 1. Recall finally that $g \circ f$ is continuous if f and g are continuous (see Exercise 7.5).

To prove the cosine is continuous at zero, we use $\cos x = \sqrt[2]{1 - \sin^2 x}$, by manipulating a familiar trigonometric identity. Admittedly this is not true everywhere (else the cosine would always be positive), but it is true on, say, $(-\pi/2, \pi/2)$ which contains 0. This is good enough to substitute $\sqrt[2]{1 - \sin^2 x}$ for $\cos x$ in proving continuity (see Proposition 6.1.1 and subsequent discussion).

Observe next that $\sqrt[2]{1 - \sin^2 x}$ is a composite function: with $f(x) = 1 - \sin^2 x$ and $g(x) = \sqrt[2]{x}$, our composite is exactly $g \circ f$. To apply the composition theorem at $x = 0$ we must check that f is continuous at 0 and that g is continuous at $f(0)$. Well, $f(0) = 1$ where the square root is continuous, taking care of g.

What about continuity of f at 0? Note $f(x) = 1 - \sin^2 x$ is *itself* a composite function: let $h(x) = 1 - x^2$ and $j(x) = \sin x$. Then $f = h \circ j$. Verify this, and catch your breath.

7.44:

So to get f continuous at 0, we need j continuous at 0 and h continuous at $j(0) = 0$. Well? If so, are we done? (Yes.)

7.45:

Proposition 7.5.2 *The cosine function is continuous at 0.*

Proof. Observe that $1 - \sin^2 x$ is continuous at $x = 0$ since it is the composition of the sine function (continuous at 0) and a polynomial (continuous everywhere). Then $\sqrt[2]{1 - \sin^2 x}$ is continuous at 0, since it is again a composition, this time of $1 - \sin^2 x$ with the square root, which is continuous at $1 - \sin^2 0 = 1$. But on an open interval surrounding 0, $\cos x = \sqrt[2]{1 - \sin^2 x}$, so the cosine is continuous at 0.

The exercises to follow provide some practice in using our theorems to build more and more complicated functions that we know are continuous.

7.5.6 Exercises

7.46: Show that the tangent function is continuous.

7.47: Show in two ways that the cotangent is continuous. Repeat for the secant.

7.48: Determine where $f(x) = \cos(x^2)$ is continuous, and prove it.

7.49: Repeat for f defined by $f(x) = \sqrt[2]{\cos(x^2)}$.

7.50: Repeat for $f(x) = (\sqrt[2]{\cos(x^2)})^3$.

7.6 Logs and Exponential Functions (Worse Trouble)

Surely logs and exponential functions should be at least as bad as trig functions. If our definition of trigonometric functions was a little insecure, what of 2^x (really? $2^{\sqrt{x}}$? 2^π? π^π?)? Also, exponentials and logs should be bad because they include roots (since, for example, $\sqrt[2]{x} = x^{\frac{1}{2}} = e^{\ln x^{\frac{1}{2}}} = e^{\frac{1}{2} \cdot \ln x}$) and we had enough trouble with just the square root previously.

The usual approach is to define the natural log function ("ln") via an integral, to define the exponential function ("exp") as its inverse function, and then to prove various properties which show that the natural log is behaving like a logarithm (e.g., $\ln a \cdot b = \ln a + \ln b$) and that exp behaves like powers of a certain number e. Next get logs and powers of other bases from these. Finally, show ln is continuous, and prove a general result about the continuity of an inverse function of a continuous function, thereby deducing continuity of exp.

This is quite an ambitious program. We'll be content with showing that ln is continuous at a single point, and leaving the rest for future courses. Here's the basic definition, relying on the assumption that $1/x$ is integrable on $(0, \infty)$.

Definition 7.6.1 *Define a function* ln *(the natural logarithm) by*

$$\ln x = \int_1^x \frac{1}{t}\, dt, \quad x > 0.$$

7.6.1 Continuity of ln

We show this "ln" function is continuous, via limits. Pictures will help with the technical details, and remember integral as "area under the curve."

Proof Discovery

Of course "continuous" means continuous on its domain, so we must show that ln is continuous at each positive number a. We will content ourselves with $a = 5$. So let $\epsilon > 0$ be arbitrary. We need $\delta > 0$ so that if $0 < |x - 5| < \delta$ then $|\ln x - \ln 5| < \epsilon$. This last inequality is

$$\int_1^x \frac{1}{t}\, dt - \int_1^5 \frac{1}{t}\, dt < \epsilon.$$

Fact 1: for integrable f and any u, v, and w,

(7.5) $$\int_u^v f(t)\, dt + \int_v^w f(t)\, dt = \int_u^w f(t)\, dt.$$

Applying this with u set to 1, v to 5, and w to x, and subtracting, we get

(7.6) $$\int_1^x \frac{1}{t}\, dt - \int_1^5 \frac{1}{t}\, dt = \int_5^x \frac{1}{t}\, dt,$$

so it is this last integral we force to be less than ϵ by forcing x δ-close to 5.

Fact 2: if f is continuous on $[c, d]$, M is the maximum value and m the minimum of f on $[c, d]$, then

$$m \cdot (d - c) \le \int_c^d f(t)\, dt \le M \cdot (d - c).$$

Recall that since f is continuous, it has a maximum and minimum on any closed interval by the Maximum Theorem (Section 4.2.3). So we can control the size of the integral in (7.6) in terms of $x - a$ (the difference of the endpoints) using M and m for the $\frac{1}{t}$ function on the relevant interval. Of course, M and m depend on x, since we must consider $\frac{1}{t}$ on all of 5 to x. We use the standard trick to ensure x is within 1 of 5: take $\delta \leq 1$ irrespective of ϵ or later requirements. Then no matter what x is, the value t (to be inserted into $\frac{1}{t}$) satisfies $4 = 5 - 1 \leq t \leq 5 + 1 = 6$. Think about the $\frac{1}{t}$ function : what value in $[4, 6]$ is gives the maximum value?

7.51:

So if $|x - 5| < \delta \leq 1$, $4 \leq x$, and M for the interval $[5, x]$ is no larger than $1/4$, and

$$\int_a^x \frac{1}{t}\, dt \leq M \cdot (x - a)$$

$$\leq \frac{1}{4}(x - 5), \qquad x \in (5 - \delta, 5 + \delta).$$

Unfortunately[5] there's a potential problem with $x - 5$. What's the problem?

7.52:

The notation $\int_5^x \frac{1}{t}\, dt$ is fine even if $x < 5$ (Fact 3: if $d < c$ we define $\int_c^d f$ to be $-\int_d^c f$) but to write "the interval $[a, x]$" is bad and $\frac{1}{4}(x - a)$, negative, won't be part of an upper bound on the integral. If $x < a$, use the fact about integrals that $\int_c^d f(t)\, dt = -\int_d^c f(t)\, dt$. Then

$$\left| \int_5^x \frac{1}{t}\, dt \right| = \left| -\int_x^5 \frac{1}{t}\, dt \right|$$

$$= \left| \int_x^5 \frac{1}{t}\, dt \right|$$

$$\leq |M \cdot (5 - x)|$$

$$= M \cdot |x - 5|$$

$$\leq \frac{1}{4}|x - 5|, \qquad x \in (5 - \delta, 5 + \delta),$$

where M is the maximum of $\frac{1}{t}$ on $[x, 5]$ if $x < 5$. Justify each of the equalities or inequalities in the chain above.

---

[5]Proof discovery frequently includes this word.

7.53:

Combining our results for the two cases yields

(7.7)
$$\left| \int_5^x \frac{1}{t}\, dt \right| \le \frac{1}{4}|x - 5|, \quad x \in (5 - \delta, 5 + \delta).$$

What are we doing all this for? Oh, yes, the goal was to force $\left| \int_5^x \frac{1}{t}\, dt \right|$ less than ϵ by forcing $|x - 5| < \delta$ (our goal is the choice of δ). What choice of δ suggests itself on the basis of the above inequality? But then remember $\delta < 1$ was required above, and adjust accordingly.

7.54:

Write up the proof, quantifiers and all, and showing your δ works.

7.55:

Remark
One moral: the proof discovery phase is usually just as imperfect and disorganized as was shown. You will frequently write down something not quite right, and have to backtrack and fix it. Write things down, though, because you can't fix what you were too timid to write down in the first place. Your goal should not be to write down a perfect, clean proof on your first try (completely unrealistic), but to capture your good ideas and then fix any technical problems afterwards (the fixing is a requirement, alas).
End Remark

The above proof was unpleasantly technical, but the exercises to follow are (with one exception) mostly easy, because our library of continuity theorems do most of the work. We pretend henceforth that we showed ln is continuous on its domain.

7.6.2 Exercises

7.56: Prove that the function f defined by $f(x) = \frac{1}{\ln x}$ is continuous.

7.57: Prove that $g(x) = \ln(\sin^2 x)$ is continuous. Where, exactly?

7.58: Suppose h is a function continuous on $\mathbf{R}$ and strictly positive there. Prove that the function j defined by $j(x) = \ln(h(x))$ is continuous on $\mathbf{R}$.

7.59: (Challenge Problem) Prove ln is continuous at each point of its domain.

7.60: (Challenge Problem) Suppose that f is any function continuous on $\mathbf{R}$, and define a new function F by $F(x) = \int_1^x f(t)\,dt$. Assume this definition makes sense (for example, assuming that any continuous function is integrable would be enough). Imitate the proof that ln is continuous to prove that F is continuous (just at 5, say). What facts are you using about integrals? About continuous functions?

8
Derivatives

Our goal (recall that we ignore applications) is the definition of the derivative, results like the derivative of the sum and the derivatives of familiar functions, and some important theorems. All our previous work on limits has done a lot of the starting work for us.

8.1   General Derivative Theorems

We begin with the definition.

Definition 8.1.1 *Let f be a function defined in an open interval about a point a. The derivative of f at a, denoted f'(a), is*

$$f'(a) = \lim_{x \to a} \frac{f(x) - f(a)}{x - a},$$

supposing this limit exists.

The expression occurring inside the limit is the "difference quotient."

Aside

   First we have to get a subtle point out of the way. It is often said that the derivative gives the slope of the tangent line. But except for the circle, there isn't a geometrically preexisting tangent line out there whose slope we capture. The tangent line to a function at a point is not an object that is already defined; the tangent line is what we are in the process of defining when we agree to make its slope the above limit. Tested in any way (e.g., does it recapture the tangent line for the circle?) this behaves beautifully. But the derivative as slope of tangent line is a *choice of definition*, not merely a convenient computational device. With the slope, we can get the equation of the tangent line to the graph of f at $(a, f(a))$, surely.

8.1:

End Aside

The first question is, do our product, sum, and so on limit theorems turn immediately into product, sum, and so on derivative theorems? Not directly. The limit we are taking is not of f, but of some function built from f, namely,

$$\Delta(x) = \frac{f(x) - f(a)}{x - a}.$$

So we'll have to go step by step: for example, is $(f+g)'(a) = f'(a) + g'(a)$?

8.2:

Even easier is that the derivative[1] of a constant times f is the constant times the derivative of f.

8.3:

We seem to have begun on the "general theorems" track ("derivatives of familiar functions" coming soon); if so, what's the derivative of a product? First guess: the derivative of the product is the product of the derivatives, reasonable because that is the way limits work, and because it generalizes the true result for sums. But it's false, and we detour for a counterexample.

Via our limit theorems, there are a couple of derivatives we can compute pretty easily. Consider f defined by $f(x) = x$, and the point $a = 1$.[2] The derivative of f at 1, supposing it exists, is

(8.1)
$$\lim_{x \to 1} \frac{x-1}{x-1}.$$

Surely this limit is 1, yielding $f'(1) = 1$; provide a reason, from one of our limit results, for this intuitively appealing conclusion.

---

[1] "Derivative" means derivative at a point for now, but "derivative" is another word, like "continuous," used several ways.

[2] In fact, the result holds for any a.

8.4:

Now evaluate the derivative of $g(x) = x^2$ at $a = 1$ in two ways, directly via the limit (factor) and then using (8.1) and the hoped-for product rule.

8.5:

Contradiction: "the limit of the product is the product of the limits" is false in general. In fact, more examples show this almost never works.

You probably knew that; we need the real "product rule" $(fg)'(x) = f(x)g'(x) + f'(x)g(x)$. The key *idea* in the proof, to supplement sensible manipulation of proof techniques, is to write

$$\lim_{x \to a} \frac{f(x) \cdot g(x) - f(a) \cdot g(a)}{x - a},$$

as

$$\lim_{x \to a} \frac{f(x) \cdot g(x) - f(x) \cdot g(a) + f(x) \cdot g(a) - f(a) \cdot g(a)}{x - a}.$$

(This is the "mathematician's trick" of adding and subtracting the same quantity.) Now break this at the plus sign and use limit rules and algebra. Write up the proof, with quantifiers for f, g, and a, please.

8.6:

Check your proof against the Hint, because there is a standard form for such limit arguments. But there's still trouble. Suppose we had to establish first the existence of all the component limits before we started using rules about sums and products. There's no trouble with $\lim_{x \to a} g(a)$ because a is a constant, so $g(a)$ is a constant, and the limit of a constant is that constant. Fine. But why does $\lim_{x \to a} f(x) = f(a)$, or even exist? This is continuity of f at a; we assumed that a different f-related function (the difference quotient) had a limit at a. Oops.

Two responses spring to mind. First, the product rule is right, you say. But a true result can have many incorrect proofs.[3] And this does beg the question of how you "know" the product rule is true if you haven't proved it.

A second reason is better: you have learned that if a function has a derivative at a point then it is continuous at that point. This has the

---

[3] "A duck is a mammal. Any mammal has feathers. Therefore a duck has feathers." In mathematics two wrongs can make a right, but don't count on it.

advantage of being true, although we haven't proved it to this point. We will do so very shortly, thus finishing the proof of the product rule as a side effect. Basically, we are in the situation in which one of our steps, although correct, requires more justification than was at first obvious. This happens all the time when you are trying to prove things; it is, in fact, where lemmas come from, since in the discovery of a proof a step requiring more difficult justification is often pulled out and proved as a lemma in the proof presentation.

A third, subtle reason is this: Observe that our assumptions for the claim are completely symmetric in f and g (f and g each have a derivative at a). Further, the concluding equation is symmetric in f and g ("derivative of f times g plus derivative of g times f"). Our *proof* seems to introduce an asymmetry: if f is continuous we never need g continuous (because we need only the limit of the constant $g(a)$). Weird. Why does f need to be continuous when g does not?

Rearrange the algebra so the continuity of g seems to be required while that of f is not.

8.7:

There are some theorems like the (just barely possible) "if f and g are differentiable at a, and one of them is continuous at a, then … ." But we will soon show that differentiability implies continuity and thus banish all these difficulties.

8.1.1 Exercises

8.8: Assuming the product rule holds, compute $f'(1)$ if $f(x) = x^3$ using only results from this text.

8.9: Another way to get the derivative of the cube uses that $x^3 - a^3 = (x-a)(x^2 + x \cdot a + a^2)$. Check this, and use it to evaluate the derivative of the cube function at a general point a (Proposition 6.1.1 helps).

8.10: Assuming again differentiability implies continuity, prove the correct result about the derivative of $\frac{1}{f}$ at a point at which f has a derivative.

8.11: Still assuming differentiability implies continuity, prove the correct result for the derivative of a quotient in terms of quotients of the derivatives.

8.2 Continuity Meets Differentiability

To get "differentiability implies continuity" isn't hard. It helps to get continuity from differentiability by showing $\lim_{x \to a} f(x) - f(a) = 0$ instead of $\lim_{x \to a} f(x) = f(a)$ (Section 6.7 gives the equivalence). This, plus one small idea, works. Try it.

Theorem 8.2.1 *If f is differentiable at the point a, then f is continuous at a.*

8.12:

We're not quite done with continuity and differentiability (compared). Could it be that continuity at a point implies differentiability, so the notions are somehow equivalent? We can present this question in terms of the following grid:

	continuous	not continuous
differentiable	?	?
not differentiable	?	?

From the result above no function lives in the "northeast" corner: it is impossible to be differentiable and not continuous. We are asking now about the "southwest" corner, but we may as well deal with all of them.

The others are easy. We know from a previous computation that $f(x) = x$ is differentiable at $a = 1$ and thus continuous there, filling one diagonal position. Similarly, we know that $\sin \frac{1}{x}$ is not continuous at 0 (we have simpler examples if you prefer) and therefore is not differentiable at 0, filling another. So only one position remains in doubt, and we are really in pursuit of a function continuous but not differentiable.

Find one: intuitions about "don't have to pick up your pencil" and "unique tangent line" may help.

8.13:

Here's a standard, computationally simple, example: $f(x) = |x|$ at $a = 0$. First, prove continuity at 0.

8.14:

Next, show this function does not have a derivative at 0, i.e., that a certain difference quotient does not have a limit at 0. Consider first examples corresponding to $x > 0$ (that is, the points relevant to the "right-hand limit" at 0 — see Exercise 2.22). Show that the right-hand limit exists. Compute a few sample numerical values for the difference quotient, and/or graph the function.

8.15:

Find (with graph) the left-hand limit for the difference quotient. Since the left- and right-hand limits are not equal, the limit (as a whole) does not exist. Examine the graph of the difference quotient for all nonzero x.

8.16:

So the grid is filled as follows:

	continuous	not continuous
differentiable	x	impossible
not differentiable	$|x|$	$\sin\frac{1}{x}$

Our proof of Theorem 8.2.1 is done, and thus so is the proof of the product theorem, and through it the reciprocal and quotient theorems.

8.3   The Monster, Weakened

With derivatives of sums, products, and quotients, we are still missing a standard tool. What is it?

8.17:

As with limits and composition of functions, so with derivatives and composition — the chain rule is harder than, say, the product rule. Here's a straightforward try, ignoring temporarily details like whether the steps can be justified or not:

$$
\begin{aligned}
\lim_{x \to a} \frac{g(f(x)) - g(f(a))}{x - a} &= \lim_{x \to a} \frac{g(f(x)) - g(f(a))}{f(x) - f(a)} \cdot \frac{f(x) - f(a)}{x - a} \\
&= \lim_{x \to a} \frac{g(f(x)) - g(f(a))}{f(x) - f(a)} \cdot \lim_{x \to a} \frac{f(x) - f(a)}{x - a} \\
&= g'(f(a)) \cdot f'(a).
\end{aligned}
$$

Now we worry: any problems above? The last equality has an easy part: $f'(a)$ is the limit claimed, assuming f is differentiable at a. (We probably needed this hypothesis all along.) The second equality is right (limit product rule) if each of the constituent limits exists.

First problem: the last inequality claims that $g'(f(a))$ is equal to a certain limit *different* from the defining limit for $g'(f(a))$. Is

$$
\lim_{x \to a} \frac{g(f(x)) - g(f(a))}{f(x) - f(a)} \overset{?}{=} \lim_{y \to f(a)} \frac{g(y) - g(f(a))}{y - f(a)}?
$$

Indeed, is g even differentiable at $f(a)$ (oops — another hypothesis we clearly needed)? Assume so; does the limit on the left exist, and is it equal to $g'(f(a))$.

Intuitively, it might. As x gets close to a, $f(x)$ gets close to $f(a)$. Why?

8.18:

So the $f(x)$'s are "y's" getting close to $f(a)$. As y gets close to $f(a)$, $\dfrac{g(y) - g(f(a))}{y - f(a)}$ gets close to $g'(f(a))$. So for the particular y's which are $f(x)$'s, $\dfrac{g(f(x)) - g(f(a))}{f(x) - f(a)}$ gets close to $g'(f(a))$. This, turned into a limit statement, is exactly what we need.

This is only intuition; can it be turned into an argument? No: consider f a constant function, $f(x) = c$ for all x. Then $f(x) = f(a)$ for all x, so

$$(8.2) \qquad \lim_{x \to a} \frac{g(f(x)) - g(f(a))}{f(x) - f(a)}$$

is trouble. Even for other f, $f(x) = f(a)$ might happen often (recall $\sin \frac{1}{x}$ and its relatives). So the first step, where we break up the limit defining the derivative of $g \circ f$ at a, is illegal, since the product of quotients may be undefined (infinitely often) though the original quotient was defined. Oops.

The proof fails, but the result might survive. The quotient in (8.2) is suspect if $f(x) = f(a)$, but for such an x the original quotient

$$(8.3) \qquad \frac{g(f(x)) - g(f(a))}{x - a}$$

is 0. Since in such a case the quotient $\dfrac{f(x) - f(a)}{x - a}$ is also zero, the value of the quotient in (8.2) seems irrelevant. Perhaps there's a chance.

The good result can be proved, via a non-quotient approach to the derivative. We'll be content with a weaker result, usually strong enough.

Theorem 8.3.1 (Restricted Chain Rule) *Let f and g be real-valued functions such that $g \circ f$ is defined in an open interval containing a, suppose that f is differentiable at a, and that g is differentiable at $f(a)$. Assume also that there is an open interval (b, c) containing a so that, for all x in (b, c), $f(x) = f(a)$ only if $x = a$. Then $g \circ f$ is differentiable at a, and $(g \circ f)'(a) = g'(f(a)) \cdot f'(a)$.*

Proof. The sequence of equalities above can be saved if

$$(8.4) \qquad \lim_{x \to a} \frac{g(f(x)) - g(f(a))}{f(x) - f(a)} = \lim_{y \to f(a)} \frac{g(y) - g(f(a))}{y - f(a)}.$$

(Note: we must show the left-hand side exists *and* is equal to the right-hand side.) We use an ϵ-δ argument; recall the right-hand limit above is assumed to exist.

Let $\epsilon > 0$ be given. Since the limit defining $g'(f(a))$ exists, there is $\gamma > 0$ so that, for all y, $0 < |y - f(a)| < \gamma$ implies $\left| \dfrac{g(y) - g(f(a))}{y - f(a)} - g'(f(a)) \right| < \epsilon.$

Now, since f is continuous at a (because differentiable there), there is $\delta_1 > 0$ so $|x - a| < \delta_1$ implies $|f(x) - f(a)| < \gamma$.

Also, there is $\delta_2 > 0$ such that if $|x - a| < \delta_2$ then $f(x) = f(a)$ implies $x = a$. Now let $\delta = \min(\delta_1, \delta_2)$. We claim that δ is as required for ϵ.

To show this, suppose x is arbitrary satisfying $0 < |x - a| < \delta$. Observe then that $f(x) \neq f(a)$, since $\delta \leq \delta_2$. Further, $|f(x) - f(a)| < \gamma$ since $\delta \leq \delta_1$. But for any y satisfying $0 < |y - f(a)| < \gamma$, in particular, for our $f(x)$,

$$\left| \frac{g(f(x)) - g(f(a))}{f(x) - f(a)} - g'(f(a)) \right| < \epsilon.$$

By the template for "$\forall x$," we have (8.4).

A reference for the proof of the full chain rule (which we use henceforth without comment) is [4].

8.3.1 Exercises

8.19: Verify that $f(x) = 2x^3 + 1$ and $g(x) = x^4 + 7$ satisfy the conditions of Theorem 8.3.1 at $a = 3$, and compute the derivative of $g \circ f$.

8.20: Repeat for $f(x) = \sin x$ and $g(x) = x^4 + 7$, assuming the derivative of the sine at a is $\cos a$.

8.4   Polynomial and Rational Derivatives

Clearly, all we need is the derivatives of the powers of x. Why?

8.21:

Also we have the derivatives of x and x^2 (yes, at $a = 1$; generalization is easy). How about higher powers? One approach relies on the binomial theorem, which lets you multiply out $(x + h)^n$, and so tackle the difference quotient for high powers of x. (See the Exercises; we used a similar idea in Exercise 8.9.)

Instead, we use induction; Exercise 8.8, the derivative of x^3 based on x^2, gives the clue. Here are the results; prove them, and the derivative of a constant function too.

Lemma 8.4.1 *Let $f(x) = x$. Then the derivative of f at any a is 1.*

Proposition 8.4.2 *Let $f_n(x) = x^n$ for n in $\mathbf{N}$. Then the derivative of f_n at any a is $n \cdot a^{n-1}$.*

8.22:

Fussy question: do we have the derivative of an arbitrary polynomial?

8.23:

There are two solutions. We could prove a "derivative of the sum of arbitrary length" theorem. Alternatively, we might *use* our theorem about the limit of such sums (see Exercise 7.30): with some algebra, the derivative turns into a limit of such a sum.

8.24:

We are done with derivatives of polynomials and rational functions — polynomials via sums and constant multiple results, rational functions via the quotient rule. Polynomials are differentiable everywhere; rational functions are differentiable everywhere defined. Nothing more to say.

8.4.1 Exercises

8.25: If f, g, and h are each differentiable at a, find a formula in terms of f, g, h, f', g', and h' for the derivative of $f \cdot g \cdot h$ at a. First, cite the product rule repeatedly (easy), then work via the definition and by imitating the product rule proof.

8.26: Suppose f is differentiable at a, g differentiable at $f(a)$, and, for all x, $(g \circ f)(x) = 1$. Derive what will be a useful relationship between the derivatives of f and g using the chain rule. You may assume that the restricted chain rule is applicable. Repeat if, for all x, $(g \circ f)(x) = x$.

8.27: Here we prove a weak version of the binomial theorem. Prove, by induction, that for any n in $\mathbf{N}$, x and h in $\mathbf{R}$, $(x + h)^n = x^n + nx^{n-1}h + C(x, h, n)$, where $C(x, h, n)$ is a sum of $n - 1$ terms, each a product of x's and h's, and each with h to the power at least 2.

8.28: Use the previous problem to evaluate, via the definition, the derivative of f_n defined by $f_n(x) = x^n$. It helps to use the result of Section 6.7 to allow the alternate definition $f'(x) = \lim_{h \to 0} \dfrac{f(x + h) - f(x)}{h}$.

8.5   Derivatives of Trigonometric Functions

For these derivatives, it is convenient to change notation. Using the result
of Section 6.7, we have that

$$f'(a) = \lim_{h \to 0} \frac{f(a+h) - f(a)}{h}.$$

(So if the usual derivative limit exists, so does the right-hand side above,
and they are equal (and vice versa).)

Recall that *continuity* of trig functions (amazingly) came down to two
limits. Why?

8.29:

What about derivatives? Consider the derivative limit for the sine at
some point a. Use the notation above, and some standard trig identities,
to show this derivative limit exists, and can be evaluated, if we know two
limits involving trig functions.

8.30:

Observe that $\lim_{h \to 0} \frac{\sin h - \sin 0}{h}$ is exactly the limit for the derivative
of the sine at 0; from a graph, and using derivative as slope of tangent line,
guess the value.

8.31:

Since the derivative of the sine at a is $\cos a$, the other limit is what?

8.32:

We now justify all this intuition.

8.5.1   Useful Trigonometric Limits

Our evaluation techniques have the same difficulties discussed in Section
7.5.2; we won't rehearse them here. But based on the following standard

picture, we can prove what we want:

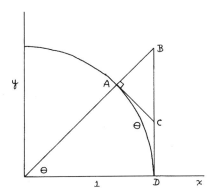

Suppose first θ is in the interval $(0, \pi/2)$. Recall also θ is the radian measure of the arc AD. Since the length of this arc is less than the sum of $|AC|$ and $|CD|$, $\theta < |AC| + |CD|$. Since AC is a leg of a right triangle with hypotenuse BC, we have $|AC| < |BC|$. Then $|AC| + |CD| < |BC| + |CD|$. Since $|BC| + |CD| = \tan\theta$, we get $\theta < \tan\theta$. This yields the left-hand side of

(8.5) $$\cos\theta < \frac{\sin\theta}{\theta} < 1, \qquad 0 < \theta < \pi/2.$$

The right-hand side was proved in the course of proving that the sine is continuous at 0 (see Section 7.5.1). Now consider some angle α in the range $(-\pi/2, 0)$; then $-\alpha \in (0, \pi/2)$. So inserting into (8.5), we get

$$\cos(-\alpha) < \frac{\sin(-\alpha)}{-\alpha} < 1,$$

and so, using $\cos(-\alpha) = \cos\alpha$ and so on,

(8.6) $$\cos\theta < \frac{\sin\theta}{\theta} < 1, \qquad \theta \in (-\pi/2, \pi/2), \theta \neq 0.$$

Recall our goal, $\lim_{\theta \to 0} \frac{\sin\theta}{\theta}$; what comes next?

8.33:

Using $\lim_{\theta \to 0} \cos\theta = \cos 0 = 1$ (Section 7.5.5), and the Squeeze Theorem (Section 6.6) we get

(8.7) $$\lim_{\theta \to 0} \frac{\sin\theta}{\theta} = 1.$$

The other limit we need comes from this one and previous theorems. Note that

$$\lim_{\theta \to 0} \frac{(\cos\theta - 1)}{\theta} = \lim_{\theta \to 0} \frac{(\cos\theta - 1)(\cos\theta + 1)}{\theta(\cos\theta + 1)},$$

and use trig identities and the new limit to finish.

8.34:

Proposition 8.5.1 *The derivative of the sine at any point a is* $\cos a$.

Prove the right result for the cosine as well; the other trig derivatives follow.

Proposition 8.5.2 *The derivative of the cosine function at any point a is* $-\sin a$.

8.35:

8.5.2   Exercises

8.36: There's a standard exercise to practice these limits, included here mostly to make sure you aren't doing something horrible, and to critique the usual solutions. If you aren't scared off yet, here's the problem: compute $\lim_{x \to 0} \frac{\sin 3x}{5x}$.

8.37: While you are at it, compute $\lim_{x \to 0} \frac{\sin 3x}{5x^2}$.

8.6   The Derivative Function

We detour to consider another meaning of the word "derivative." To date this has meant only the derivative at a point, so $f'(a)$ was the limit of the appropriate difference quotient at a. You know, and the notation hints at, more: "$f'(a)$" looks like some function whose name is "f'" evaluated at some point "a." We haven't had the function yet; we could have called the limit "b" just as well.

But the notation is suggestive for a reason. If the derivative of f at the point 1 were called b, it would be inconvenient to discuss the derivative of f at 2, let alone the derivative at lots of points. The device is this: we create a function, called f', whose value at each point a is the derivative of f at a. We really assemble a function from a long list of ordered pairs, where the first element of a pair is some point a and the second element is the derivative of f at a.

Aside

It may be time to fix some bad notational habits. If you've written "the function $f(x)$," realize that this is notationally nonsense, since the name of the function is f while $f(x)$ is some value of the function at x. Often "knowing what you mean" is good enough; but if you aren't really sure about the difference between the function and the value of the function, this habit and its notational abuse are deadly. It helps to write things carefully from now on.

End Aside

Sometimes we separate the two uses of "derivative" by language such as "derivative at a point" (some $f'(a)$) and "derivative function" (f'). If we are being notationally correct this distinction is unnecessary (because it is redundant, if potentially helpful nonetheless): "the derivative function f'" is redundant because f' couldn't mean derivative at a point, since f' is a function and not a number. Just for practice, make up the other redundant but correct phrase and explain why it is redundant.

8.38:

Give the two awful, unsayable, self-contradictory, phrases.

8.39:

There's yet another confusing factor. We're more used to "the function f defined by $f(x) = x^2$" than "the function f defined by $f(a) = a^2$," although they do mean the same thing. But try taking $f'(x)$ using the usual difference quotient:

$$f'(x) = \lim_{x \to x} \frac{f(x) - f(x)}{x - x}.$$

Trouble again: "x" is standing for two different things, one the *fixed* point at which we are taking the difference quotient (formerly a) the other a *different* variable "x" doing the "approaching."

One approach is to change the basic notion of derivative to be

$$f'(x) = \lim_{h \to 0} \frac{f(x + h) - f(x)}{h},$$

producing $f'(x)$ (the value) directly. Basically, though, you just have to stay alert and cope.

With the derivative as function, we can consider old theorems in new clothes. For example, if f is defined by $f(x) = x^n$ then f' is defined by $f'(x) = nx^{n-1}$ (this is the function version of Proposition 8.4.2). Or, the derivative of the sine is the cosine (compare to Proposition 8.5.1). We can also phrase questions we can't answer yet, such as "if f' is a positive function, is f increasing?" From now on, we leave you to sort out the various uses of "derivative."

8.6.1 *Exercise*

8.40: Compute the equation of the tangent line to f at an arbitrary point, where f is the function defined by $f(x) = x^3 + 2x$.

8.7   Derivatives of Rational Powers

This section might be surprising; we only got *continuity* for $\sqrt{x}$ at one point, and only indicated what to do for other roots. We'll only do a little (by brute force) here, and discuss briefly the route to use for the general case.

Begin with a simple case (the square root) to see what the task is. Where does $\sqrt{x}$ have a derivative? The graph (viewing derivative as slope of tangent line) says at all $x > 0$. So we face, for some $a > 0$, the limit

$$\lim_{x \to a} \frac{\sqrt{x} - \sqrt{a}}{x - a}.$$

Here's a common approach. You are supposed to be eager to "rationalize the numerator" by multiplying top and bottom by $\sqrt{x} + \sqrt{a}$, simplifying, and then realizing the limit just falls out. Do those computations, assuming you can't contain your enthusiasm for rationalizing the numerator a moment longer. Note continuity of $\sqrt{x}$ is used.

8.41:

This approach is unbelievably frustrating. This business of "rationalizing" things was originally introduced solely to "simplify," usually a time consuming and stupid task, and always used on the denominator at that. So why would one rationalize the numerator? Further, since one grudgingly admits it works, how would anybody think of it in the first place? Finally, where's even the derivative of the cube root to come from?

These objections are perfectly fair. There is another approach; the underlying idea behind "rationalizing" can be generalized (see Exercise 8.47) to

$$x - a = (\sqrt[n]{x} - \sqrt[n]{a}) \cdot ((\sqrt[n]{x})^{n-1} + (\sqrt[n]{x})^{n-2}(\sqrt[n]{a})^1 + \ldots + (\sqrt[n]{x})^1(\sqrt[n]{a})^{n-2} + (\sqrt[n]{a})^{n-1}).$$

Use this formula with $n = 3$ to manipulate the limit for the cube root derivative to generate the expected formula (assume $\sqrt[3]{x}$ is continuous).

8.42:

Do the task for general n (hint: the sum has n terms).

8.43:

We record these results as a theorem. Did you catch trouble at $x = 0$?

Theorem 8.7.1 *For each n in **N**, let $f_n(x) = \sqrt[n]{x}$ for all suitable x. Then if n is even, $f'_n(a) = \dfrac{1}{n(\sqrt[n]{a})^{n-1}}$ for all $a > 0$, with the same equation for all $a \neq 0$ if n is odd.*

Since $x^{p/q} = \sqrt[q]{x^p}$, a formula for the derivative of any rational power follows using the chain rule.

8.44:

Remarks
(1) A more common approach to derivatives of rational functions is "implicit derivatives," computationally simple. But justification of implicit derivatives is quite nontrivial. (Indeed, are you prepared to say, exactly, what an implicitly defined function is? If not, what are you doing taking the derivative of one?)
(2) One can get the right formula easily if only (somehow) it were known that roots were differentiable. See Exercise 8.46 for this approach.
(3) The best way to get derivatives of root functions is by theorems about inverse functions (as we might have done in Section 7.6 for an approach to continuity). The result is the following.

Theorem 8.7.2 *Let f be a function with a derivative at each point in its domain and which is in addition invertible, say with inverse g. Then for any a in the domain of g, if $f'(g(a)) \neq 0$, then $g'(a) = \dfrac{1}{f'(g(a))}$.*

If $f(x) = x^n$, and $g(x) = \sqrt[n]{x}$, this gives the right derivative for the root. But while the proof of this result is not beyond us, it is technically annoying enough to omit. (See the next section for further discussion.)
(4) The formula we got for the derivative of roots yields, via the chain rule, $\dfrac{dx^{p/q}}{dx} = \dfrac{p}{q} \cdot (\sqrt[q]{x})^{p-q}$. This is correct, but conceptually much worse than $\dfrac{dx^{p/q}}{dx} = \dfrac{p}{q} \cdot x^{p/q-1}$. Why?

8.45:

The "power rule" for integer powers is clean and simple and works just as well for rational functions; there really is only one, very general, power rule. A better approach, sketched in the next section, yields this better result in a cleaner way.

8.7.1   Exercises

8.46: Assume for the moment that the square root function is in fact differentiable. Call the square root function f; we know that $f(x)^2 = x$ for all $x > 0$. Compute via the chain rule the derivative of $f(x)^2$; note that it is the same as the derivative of x by the equality above, and discover the derivative of f. This is essentially the "implicit derivatives" approach, and you can see why it looks simple. The hard part lies in justifying that the root function is differentiable in the first place.

8.47: For the brute force approach to the derivatives of roots, we need first to establish that, for all positive integers n and all real numbers c and d,

$$(c^n - d^n) = (c - d)(c^{n-1} + c^{n-2}d + \ldots + cd^{n-2} + d^{n-1}).$$

Prove this. We get what we used with $c = \sqrt[n]{x}$ and $d = \sqrt[n]{a}$.

8.48: Complete the proof for the derivative of the root function.

8.8   Derivatives of Exponential and Logarithmic Functions

You will be unsurprised to find these are going to be more troublesome. Since the natural log is defined in terms of an integral, we must take the derivative of an integral. If you know the Fundamental Theorem of Calculus, this point is already taken care of. Since $\ln x = \int_1^x \frac{1}{t} dt$, then the derivative at a point a is $\frac{1}{a}$. We'll return to this when we touch on integrals, and we will show what needs to be assumed to get it, but now we must simply accept it. Exponentials must be trouble because their derivatives would give derivatives of roots (see the start of Section 7.6).

The good route to the derivative of the exponential function uses the fact that exp is the inverse of ln, and the theorem about inverses from Section 8.7. Show with that with $f(x) = \ln x$ and $g(x) = \exp x$ you get the expected result.

8.49:

Another route, as in Exercise 8.46, is to assume that exp is differentiable, write $\ln(\exp(x)) = x$, take the derivative of both sides using the chain rule, and get what you want. This is akin to an "implicit derivatives" approach.

8.50:

Finally, armed with the chain rule and the derivative of exp, a clean version of the power rule is easy. Let r be any real number not 1: since $x^r = e^{\ln(x^r)} = e^{r \cdot \ln x}$, the derivative is (simplify, please):

8.51:

In fact, the theorem is good enough to get not only the derivative of exp, but the derivative of any function with an inverse and a derivative for any function for which we cut down the domain enough to produce an inverse.

More remarks about the inverse derivative theorem are in order. First, while the implicit derivative approach is hard to justify, and the "assume it's differentiable" approach is a problem, realize that they are useful for the full proof we don't give, because it is always easier to prove something about limits when you have a candidate for the limiting value, and a version of Exercise 8.46 with general f and g gives such a candidate.

Second, although we won't do a proof here, we will criticize somebody else's. (Hmmm!) Here's part of a proof one might see in an introductory calculus textbook. Recall first that if f and g are inverses, then $w = f(z)$ if and only if $z = g(w)$. Then, with $y = g(x)$ and $g(a) = b$,

$$
\begin{aligned}
g'(a) &= \lim_{x \to a} \frac{g(x) - g(a)}{x - a} \\
&= \lim_{y \to b} \frac{y - b}{f(y) - f(b)} \\
&= \lim_{y \to b} \frac{1}{\dfrac{f(y) - f(b)}{y - b}} \\
&= \frac{1}{\lim_{y \to b} \dfrac{f(y) - f(b)}{y - b}} \\
&= \frac{1}{f'(b)} \\
&= \frac{1}{f'(g(a))}.
\end{aligned}
$$

There are various steps to be justified (e.g., using "the limit of the quotient is the quotient of the limits") but the second equality is the real problem. The justification usually given is that since g is continuous at a, as $x \to a$ then $g(x) \to g(a)$, i.e., $y \to b$. Therefore, one can make the exchange of limits claimed by the second equality.

Well, this proof is a difficult and complex one to talk about. The justification for the second equality given above is not meaningful, because we have never discussed the symbol $x \to a$ in isolation at all. (It has always appeared as notation attached to a limit; while it may remind us of something, the definition of limit is really the meaningful thing.) Indeed, the justification above is almost a notational one, in which we claim something like "$x \to a$ is the same as $y \to b$" and so we can exchange the limit. Since neither of the "$\to$" symbols means anything by itself, neither does this claim.

This proof can be saved, because it *is* possible to justify (using ϵ's and δ's) the exchange of limits claimed in the second equality. But the proof is incomplete without that, and unsatisfying because the "interchange of limits" proof is at least as hard as arguing for the derivative of g directly. It doesn't seem fair to prove something by citing a harder proof, which is then not presented. We therefore leave this story for a later course.

8.8.1   Exercises

8.52: Compute the derivative of the function f defined by $f(x) = e^{\sin x^2}$. More importantly, justify your result by reference to our theorems.

8.53: To practice cutting down the domain of a function to produce an inverse, consider the usual square function s, defined on **R**. It is not invertible since it is not injective (note $s(2) = s(-2)$). Define $\hat{s}$ as follows: $\hat{s}(x) = x^2$ for all $x \geq 0$. Observe that $\hat{s}$ is the restriction of s, meaning it coincides with s on its domain, which is a (proper) subset of the domain of s. Argue graphically that $\hat{s}$ has an inverse. Now use Theorem 8.7.2 to produce a formula for the derivative of the square root function.

Warning: real life relies on alertness, not different notation, to make sure you know which of the two possible square functions is in play.

8.54: Here we work toward an inverse of the sine function and again must sacrifice domain; the standard choice for the domain of the restriction is $[-\pi/2, \pi/2]$. Show that this restriction is injective, and therefore has an inverse. What is the domain of the inverse? Its range? We call this inverse the arcsine, or inverse sine, and denote it arcsin. Is the arcsine increasing, decreasing, or neither? For which x is the following true: $\arcsin(\sin x) = x$?

8.55: Now find the derivative of the arcsine. In simplifying the result to the usual formula you will run into $\cos(\arcsin(x))$, which can be simplified by reading $\arcsin y$ as "the angle whose sine is y" and drawing a right triangle in which both this angle and y itself appear.

8.56: Produce the arctangent (the usual range is $(-\pi/2, \pi/2)$), and its derivative.

9

Theorems about the Derivative

All right, many familiar functions have familiar derivatives. Who cares, and why?

9.1:

Reasonable answers could involve the powerful and important applications of the derivative (e.g., to real-world maximum/minimum problems). But these applications are as yet unjustified by us: if the derivative of a polynomial is useful, we can compute it, but nothing so far shows that it *is* useful. We turn to results underlying (some of) the applications.

9.1 The Derivative and Extrema

We first set up the language for maximum/minimum problems and the derivative.

9.1.1 *Preliminaries*

Definition 9.1.1 *Let f be a function defined on an open interval surrounding a point a. We say a is a <u>local</u> <u>maximum</u> <u>point</u> for f if there exists some δ > 0 so that $f(a) \geq f(x)$ for all x satisfying $|x - a| < \delta$. The value $f(a)$ is called a <u>local</u> <u>maximum</u> <u>value</u> for f.*

Explore: what is the significance of $f(a) \geq f(x)$ as opposed to $f(a) > f(x)$? Draw a picture of a local maximum with another local maximum nearby, indicating the significance of "δ." If δ_1 is as needed to show a is a local maximum point, what about δ larger than δ_1? Smaller than δ_1? Write the definition in "neighborhood" terms instead of as presently written. Define local minimum point and local minimum value. Can a point a be *both* a local maximum point and local minimum point for some f?

9.2:

To bundle maxima and minima together, we call them <u>extrema</u> (plural of <u>extremum</u>) and say "local extreme point" or "local extreme value" and so on. Recall also the maximum of a function on a set (Section 4.2.1).

Definition 9.1.2 *Let f be defined on some subset of* $\mathbf{R}$ *and S a subset of the domain of f. The point b in S is a maximum point for f on S if*

$$f(b) \geq f(x), \quad \text{for all } x \in S.$$

Review, including Section 4.2.1 as needed. Give the "minimum" version.

9.3:

What's the relationship between our two "maximum" definitions? It helps to see that local maximum point requires the existence of a certain very special set S containing b such that b is a maximum point for f on S. What is the form of S? Construct f and S so that f has a maximum point on S that is not a local maximum for f. (Section 4.2.1 may help.) Then show the world can be good; produce f and S (say, a closed interval, or $\mathbf{R}$) with a maximum on S that is also a local maximum. Give also an f and an S with a local maximum point in S that is not a maximum for f on S. Finally, summarize the (rather loose) relationship between the two definitions by filling the grid with examples, or determining that a cell must be empty:

	local maximum point	not local maximum point
maximum point on set	?	?
not maximum point on set	?	?

9.4:

So the two definitions are complementary rather than coincident. But you doubtless recall from applied calculus problems (and we show next) that although the maximum of a function on a closed interval need not be a local max, and endpoints must be examined, no other points need apply.

9.1.2   The Derivative and Local Extrema

You recall from applied calculus max/min problems the importance of points where f' is zero. But step back: here are two candidates for a relationship.

i) If a is a point where $f'(a) = 0$, then a is a local extreme point for f.

ii) If a is a local extreme point for f, then $f'(a) = 0$.

Sadly, neither of these preliminary candidates is true. The cube function disposes of the first, the absolute value function the second.

9.5:

We must repair these until at least one is true.[1] But the first is just doomed: x^3 is honestly differentiable (everywhere) and "derivative zero" doesn't imply "local extremum." End of story.

One fix for the other is to consider only functions differentiable everywhere, so $|x|$ is ruled out. Another patch is more general and almost as painless:

ii') If a is a local extreme point for f, then $f'(a) = 0$ or $f'(a)$ does not exist.

It turns out that this is both true and useful.

Aside

Why useful? Suppose you have f whose derivative you can compute, and you need the local extrema of f. To start, you aren't sure there are any, and you have (presumably) infinitely many points, all equally good candidates. Life is too short to examine each according to the definition. With the derivative, you can find the point(s) at which f' is zero or does not exist. The set of all such points contains all local extrema (if there are any). Why?

9.6:

---

[1]This is an absolutely standard process in mathematics. Any, even very good, intuition is likely to be flawed because of unanticipated special cases or technicalities. The job (fun) is to fix things, keeping the simple spirit of the first intuition.

Often this $f'(a) = 0$ or $f'(a)$ DNE set is comparatively small. This reduces, drastically, the collection of points you have to examine.
End Aside

To prove ii') we actually prove the contrapositive. Given '$P \Rightarrow Q$,' the contrapositive is '$\neg Q \Rightarrow \neg P$' (recall "$\neg$" denotes "not"), and via truth tables (see Section 3.1) under any assignment of truth values to P and Q the implication and its contrapositive are either both true or both false.

9.7:

So to prove one is to prove the other, since they go together. This standard technique is "proof by contraposition."

The contrapositive is "If not ($f'(a) = 0$ or $f'(a)$ DNE) then a is not a local extremum of f." To manipulate this, we use the fact (truth tables again) that '$\neg(P$ or Q)' is equivalent to '$\neg P$ and $\neg Q$.' So "not ($f'(a) = 0$ or $f'(a)$ DNE)" is equivalent to "$f'(a) \neq 0$ and $f'(a)$ exists."[2] So we must prove:

ii'') If $f'(a)$ exists and $f'(a) \neq 0$ then a is not a local extreme point for f.

(If these logical manipulations are foreign to you, just take this as the goal. Eventually you'll become comfortable with this sort of thing. An intuitive version reads: To show local extrema can occur only where $f'(a) = 0$ or $f'(a)$ DNE, show that looking where $f'(a) \neq 0$ and $f'(a)$ exists is foolish since such points can't possibly be local extrema.)

Well, if $f'(a)$ exists and is not zero, there are two possibilities: $f'(a) > 0$ and $f'(a) < 0$. Consider first the case $f'(a) > 0$. Draw a "generic" picture, with derivative as slope of the tangent line, of some a where $f'(a) > 0$. Could a be, for example, a local maximum point? If so, the definition requires that for a little way on either side ("δ") of a, $f(a)$ is at least as large as any other function value. Points to one side of a seem troublesome.

9.8:

Let's prove that a is not a local maximum point (assuming $f'(a) > 0$). We must show any proposed $\delta > 0$ doesn't satisfy the local maximum point definition. Review that definition!

9.9:

---

[2]It is usual to write this in the other order, "$f'(a)$ exists and $f'(a) \neq 0$," since it makes more sense to note that it exists before talking about its value.

Suppose $\delta > 0$ is claimed to work. You must produce an x satisfying $|x - a| < \delta$ and $f(x) > f(a)$ (showing δ fails). Clearly you must somehow use $f'(a) > 0$. But $f'(a) > 0$ doesn't speak to values of f ...or does it?

9.10:

Here's the intuition. If $f'(a) > 0$, then, since $f'(a)$ is the limit of things like $\dfrac{f(x) - f(a)}{x - a}$, these terms themselves should be positive. Recall that if a limit is positive the terms must be positive, at least for x "close enough" to a (see Exercise 6.17). Well, suppose that for some x close to (but larger than) a we had $\dfrac{f(x) - f(a)}{x - a} > 0$. Why would we care?

9.11:

So such an x, if also $|x - a| < \delta$, would show that the proposed region was *not* as required for "a is a local maximum point": contradiction.

The full proof follows. Putting it all together is a little formidable, but note the ideas above are the keys.

Proposition 9.1.3 *Let f be a real-valued function of a real variable. If a is a local extreme point for f, then $f'(a) = 0$ or $f'(a)$ does not exist.*

Proof. To show that a local maximum point occurs only where $f'(a) = 0$ or $f'(a)$ DNE, we show that if $f'(a)$ exists and $f'(a) > 0$ then a is not a local maximum point. Suppose for a contradiction that $f'(a)$ exists, $f'(a) > 0$, and a is a local maximum point. There exists $\delta > 0$ so that $|x - a| < \delta$ implies $f(x) \leq f(a)$.

Since $f'(a) > 0$, we have

$$\lim_{x \to a} \frac{f(x) - f(a)}{x - a} > 0.$$

Then for g defined by $g(x) = \dfrac{f(x) - f(a)}{x - a}$ for $a \neq 0$, we are sure that $\lim_{x \to a} g(x) > 0$. Citing Exercise 6.17, there exists a punctured neighborhood of a defined by some $\delta_1 > 0$ so that for all x satisfying $0 < |x - a| < \delta_1$,

(9.1)
$$g(x) = \frac{f(x) - f(a)}{x - a} > f'(a)/2 > 0.$$

Consider now x_0 defined by $x_0 = a + \dfrac{\min\{\delta, \delta_1\}}{2}$. It is easy to verify that $x_0 > a$, that $|x_0 - a| < \delta$, and that $0 < |x_0 - a| < \delta_1$. From the last of these,

$$g(x_0) = \frac{f(x_0) - f(a)}{x_0 - a} > 0.$$

Since the fraction above is positive, and $x_0 > a$ so $x_0 - a > 0$, we have $f(x_0) - f(a) > 0$ and so $f(x_0) > f(a)$. Since $|x_0 - a| < \delta$, this contradicts the assumption on δ, and the proof by contradiction is complete.

Critique, carefully, your reading of the proof.

9.12:

Even with active reading stimulated, we must see where we are. Our goal is to prove that extreme points occur only where $f'(a) = 0$ or $f'(a)$ DNE, by showing that if $f'(a)$ exists and $f'(a) \neq 0$ then a isn't an extreme point. Did we achieve this? What part? What remains?

9.13:

The exercises to follow will let you finish, thus justifying your past approach to "unconstrained" max/min problems. (For an audience very proficient at proof, one might say that "mutatis mutandis" the other results follow, meaning, "changing the proof already done slightly you get the rest." We won't do that yet.) We'll consider extrema on a closed interval too.

9.1.3   Exercises

9.14: Case 1a: $f'(a)$ exists, $f'(a) > 0$, and we want that a is not a local *minimum* point. Via a "generic" picture, locate some bad x. Modify the partial proof of Proposition 9.1.3 for this case (no "easy to verify," please!).

9.15: The case $f'(a) < 0$ is indeed coming. But the proofs already given used a previous result (from Exercise 6.17) about "the limit is positive so the function itself is positive nearby." Unfortunately, we lack an analogous "the limit is negative so ..." result, and we need it.

Proposition 9.1.4 *Suppose f has limit $L < 0$ at the point b. Then there exists a punctured neighborhood of b on which f is bounded above by $L/2$.*

There are two approaches. One is "mutatis mutandis": go back to the original proof and make little changes as needed. But better is to *use* the previous result. We face f such that $\lim_{x \to b} f(x) = L$ and $L < 0$. Consider the function $-f$, where $-f$ is defined by $(-f)(x) = -(f(x))$ for all x. Draw f and $-f$ on the same graph. What is the limit L' of $-f$ at a? Can you prove it (yes — see Proposition 6.1.2)? Note that $L' > 0$ (for $-f$) and so, for $-f$, there is a punctured neighborhood (equivalently, δ) on which something happens for $-f$. Get from δ a punctured neighborhood as needed for f. Redraw the picture, write the proof, then check the Hint.

9.16: Tackle the remainder of the Proposition using Exercise 9.15. Namely, consider $f'(a) < 0$. There are two subcases; show in each of them, and in full detail, that a cannot be a local extreme point.

9.17: A different approach to the $f'(a) < 0$ case is to *use* the already proved $f'(a) > 0$ case as opposed to modifying its proof. Ignore Exercise 9.16 and suppose $f'(a) < 0$. Consider again $-f$. What can we say about derivatives? A local minimum point for $-f$ is a ???? for f. Write up a proof for the $f'(a) < 0$ case based on this approach. [Hint: proof by contradiction.]

9.18: To justify the usual calculus approach for "absolute" max/min on a *closed interval* $[a, b]$ (in our language, the max and min of f on $[a, b]$), we must show that an extreme point for f on the interval $[a, b]$ can only occur where f' is zero or does not exist, or at either a or b.

Well, suppose we have a maximum of f on $[a, b]$, say, at some point c neither a nor b. Must the $f'(c)$ be zero or undefined? Yes: show that such a maximum point is actually a local maximum point for f, and cite the proposition. Outline two approaches to the "minimum" version.

9.2 The Mean Value Theorem

Remark, to begin, that "mean" is used here in the sense of "average." We use also the special definition of "continuous on a closed interval" (Exercise 2.22).

Theorem 9.2.1 (Mean Value Theorem (MVT)) *Let f be continuous on the closed interval $[a, b]$ and differentiable on the open interval (a, b). Then there exists a point c in (a, b) so that*

$$f'(c) = \frac{f(b) - f(a)}{b - a}.$$

Intuitive interpretation: the expression on the right-hand side is the slope of the line through the points $(a, f(a))$ and $(b, f(b))$ (the "chord"), and we call this the average slope of f on $[a, b]$. The quantity $f'(c)$ is the slope of the tangent line at some point c strictly between a and b. Conclusion: there is some point c where the tangent line ("instantaneous") slope matches the average slope.

Draw the generic or some familiar specific pictures. Include everything.

9.19:

A standard example to give intuition views f as the position function of an object as a function of time t, $a \le t \le b$. The slope of the chord is then the average velocity of the object over the time interval (total distance traveled = final position – initial position, divided by the length of time $b - a$). The slope of the tangent line is then instantaneous velocity $f'(c)$. The statement is that at some time during the journey the instantaneous velocity exactly matched the average velocity for the whole journey. Threatened application: your average velocity on the turnpike is 70 mph, and you are ticketed by an overeducated police officer because at some time during your trip the MVT guarantees your speed (the reading on your speedometer) was exactly 70 mph, 5 miles over the limit.[3]

You may have noticed that the hypotheses are redundant: f differentiable on (a, b) guarantees f continuous on (a, b), so the only further continuity needed is of f at a and b. The given hypotheses are equivalent, and customary. Also, this theorem is what's called an "existence" theorem, guaranteeing the existence of a certain point c without any recipe for computing it in general.[4] With simple functions, c is computable (no theorem needed), but harder examples follow.

9.2.1 Exercises

9.20: Find the point or points c for f defined by $f(x) = x^2$, $a = 2$, $b = 4$.

9.21: Consider the sine function on $[0, \pi/2]$ and try computing c. Note that the MVT actually has some teeth after all.

9.2.2 Rolle's Theorem

Think about how you will prove the Mean Value Theorem.

9.22:

Lacking other inspiration, the approach below begins to make more sense. Limit ourselves first to the case $f(a) = f(b)$. Draw the picture, and write down what we want for $f'(c)$.

9.23:

Good; we have lots of experience with the special goal for $f'(c)$. In particular, if z is an extreme point for f in (a, b), it would be a local extreme point (see Exercise 9.18), and hence would have $f'(z) = 0$, as needed.

---

[3]If these numerical values are unrealistic for your driving, substitute your own.
[4]Cf. the Maximum Theorem (Section 4.2) and the IVT (Section 4.1).

Must f have an extreme point in the *open* interval (a, b)? The Maximum Theorem works for closed intervals only. But we can save things: what if there is some x in (a, b) such that $f(x) > f(a)$ (recall $f(a) = f(b)$)? Clearly neither a nor b yields a maximum of f on $[a, b]$. The Maximum Theorem gives *some* point z in $[a, b]$ yielding a maximum value,[5] and since it isn't a or b it must be in (a, b). Call it c. What about $f'(c)$ using Exercise 9.18?[6]

Of course, there may be no x in (a, b) such that $f(x) > f(a)$. There might be x, in that case, so $f(x) < f(a)$. Argue this case.

9.24:

Perhaps neither case occurs, i.e., there is no x in (a, b) such that $f(x) > f(a)$ and also no x in (a, b) such that $f(x) < f(a)$. Therefore (picture!), ...

9.25:

Observe then f is constant (identically $f(a)$). What is the derivative of a constant? Get c.

9.26:

Take a moment to write up the whole proof in an organized fashion. What you have proved is Rolle's Theorem.

Theorem 9.2.2 (Rolle's Theorem) *Let f be continuous on $[a, b]$ and differentiable on (a, b), and with $f(a) = f(b)$. Then there exists c in (a, b) so $f'(c) = 0$.*

By the way, often Rolle's Theorem appears only long enough to prove the MVT, although we will use it later.

9.2.3 Exercises

9.27: Find examples f and $[a, b]$ having exactly one Rolle's Theorem point c, exactly two, and infinitely many. Find an "infinitely many" case so the set of c is *not* a closed interval.

9.28: Find a Rolle's Theorem example in which c is not hand computable.

---

[5]What hypothesis on f do we use here? What hypothesis is unneeded yet?
[6]What hypothesis on f are we using here?

9.2.4   The Theorem Itself

Surprise: the MVT can be proved from its special case, Rolle's Theorem. Faced with a general MVT style f, we'll construct a Rolle's Theorem g from it, and so that $g'(c) = 0$ will yield the MVT conclusion for f.

We construct such a g.[7] We will know that for g and the resulting c, $g'(c) = 0$. We will want that c satisfies, for f, an equation. Find it, and then transform this equation into one with 0 on one side.

9.29:

To get this from $g'(c) = 0$, $g'(c)$ should be what?

9.30:

This is an equation involving the derivative of g (a "differential equation"). Whether these are familiar or not, it seems reasonable to make g the difference of two functions, the first having derivative f' and the second having derivative the difference quotient. A function whose derivative is f' is childishly simple. For the second term, note that the expression

$$\frac{f(b) - f(a)}{b - a}$$

is an (ugly) constant, since f, a, and b are given. There's an easy function whose derivative is 5; what?

9.31:

For the more complicated constant $\dfrac{f(b) - f(a)}{b - a}$ the good function is ...

9.32:

Putting this all together, the natural candidate for g is

$$g(x) = f(x) - x \cdot \frac{f(b) - f(a)}{b - a}.$$

Check that if $g'(c) = 0$ then c satisfies what we need for f. Check also (citing theorems) that g satisfies (*all*) the hypotheses for Rolle's Theorem.

---

[7]This construction of g is perfectly well known, if perhaps not to you. To simply announce the "magic" g encourages the reader to think it *is* magic. But given the basic idea, the construction is almost forced upon you.

9.33:

Write up the complete proof in an organized fashion. In proof writing we don't give the train of thought leading up to the proof, and simply present the results, so you *do* simply announce g and show it works.

9.34:

Remarks

Writing proofs giving just the argument (no discovery steps) may seem peculiar. Partly, a proof must *prove*, and inclusion of the intuitive idea may *convince* even if there is a subtle flaw. Some of the habit is mere custom, but one difficult for people trying to learn how to prove things.

Also, some presentations of the MVT use a slightly different function, g_1, satisfying also $g_1(a) = g_1(b) = 0$. Their g_1 is a constant plus our g (so $g' = g_1'$). There is no change in the proof either, but g_1 has a convenient geometrical interpretation, namely the vertical distance between the graph of the chord connecting $(a, f(a))$ and $(b, f(b))$ and the graph of f, at each point in the interval $[a, b]$. The condition $g(a) = 0$ gives the correct constant.

End Remarks

9.2.5 Exercises

9.35: Now find the alternative function g_1 satisfying, along with other required properties for Rolle's Theorem, $g_1(a) = 0$. Convince yourself (picture) that at each x in $[a, b]$ g_1 gives the vertical distance between the graph of the chord connecting $(a, f(a))$ and $(b, f(b))$ and the graph of f. Geometrically, one expects a maximum for g_1.

9.36: Suppose that f and g each satisfy the hypotheses of the MVT on $[a, b]$, and also $f(a) = g(a)$ and $f(b) = g(b)$. Prove that there is c in (a, b) at which $f'(c) = g'(c)$. Interpret graphically.

9.37: (Challenge Problem: Cauchy's Formula) Prove that if f and g are defined on $[a, b]$, continuous there, differentiable on (a, b), and such that $g'(x) \neq 0$ on (a, b), then there exists c in (a, b) such that

$$\frac{f(b) - f(a)}{g(b) - g(a)} = \frac{f'(c)}{g'(c)}.$$

By inserting $(b - a)$'s, one can write the left-hand side as a quotient of MVT type quotients, one for f and one for g. But Cauchy's Formula does not follow trivially from the MVT in spite of that. Why not?

9.38: (Continued) To prove Cauchy's Formula, we construct h and let Rolle's Theorem do some work. Hints: First, $g(b) \neq g(a)$, or else for some z, $g'(z) = 0$, so the proposed equation makes sense. Second, the function h we produce will fit Rolle's Theorem, not just the MVT. Remark: Cauchy's Formula is used to derive l'Hôpital's rule for derivatives.

9.3   Consequences of the Mean Value Theorem

The MVT has consequences basic for calculus and its applications (some familiar, hence "obvious"). Here's one, to answer a question at first merely paranoid, then frightening. Supposing $f'(x) = 2x$, what can we say about f? Guess number one is $f(x) = x^2$ (derivative results backwards). But $f(x) = x^2 + 5$ also works, and you probably learned to summarize matters by $f(x) = x^2 + C$, where C is any constant. Fine.

The paranoid question is, are there any *other* functions with derivative $2x$? This isn't just theoretical, because in many important applications you have the f' and want f. Think of acceleration, velocity, position problems, or, if you haven't run into those yet in calculus or physics, think of sending a rocket to the moon. We control the force produced by the engines, which turns out to be proportional to the derivative of the velocity. Knowing the derivative of velocity we get the velocity function. Knowing the velocity, i.e., the derivative of position, we get the position function from it and the astronauts arrive as desired. It would be embarrassing if they landed on Pluto because faced with $v(t) = 2t$ we used $s(t) = 2t + 6$ but really $s(t) = e^{\sin \log(755.9t^{13} + 17t)}$, another function which just happened to have derivative $2t$.

Yes, the above mess does not have derivative (anything close to) $2t$ (chain rule plus derivative formulas), but now the question is on the table. Maybe some yet uglier function has derivative $2t$, though not one of the known suspects, namely $t^2 + C$. We can't check all functions, so now what?

The following result is the basis of the negative answer to this sort of question, and is in fact a special case of it.

Proposition 9.3.1 *If f is defined on (a, b), and $f'(x) = 0$ on (a, b), then f is constant on (a, b), meaning there is a constant C such that $f(x) = C$ for all x in (a, b).*

A proof by contradiction is not very difficult.[8] Suppose there are two points x_1 and x_2 in the interval (a, b) so that $f(x_1) \neq f(x_2)$. Labeling the points so $x_1 < x_2$ doesn't hurt. Argue that f satisfies the hypotheses of the MVT on the interval (x_1, x_2). Deduce something about some c in (x_1, x_2), hence

---

[8]Actually, a proof by contradiction is not best ("aesthetically"), if easiest for those not yet comfortable with quantifiers. See Exercise 9.43 for a "better" proof.

in (a, b), and get a contradiction.

9.39:

   To obtain the constant C guaranteed by the proposition, take the function value at any point of the interval.
   The general result ("know f given f''") follows by proper use of the special case ("know f if $f' = 0$").

Corollary 9.3.2 *Suppose f and g are defined on (a, b) and $f'(x) = g'(x)$ for all x in (a, b). Then f and g differ by a constant on (a, b), meaning there exists some constant C so that $f(x) = g(x) + C$ for all x in (a, b).*

To prove this, cook up the right function and apply Proposition 9.3.1.

9.40:

Done: to know the derivative is to know the function up to a constant. (Although basically right, this intuitive statement may be pushed too far; see Exercise 9.44.)
   Another consequence of the MVT is used when graphing f based on $f' > 0$ and $f' < 0$. In traditional calculus texts sketching the graph of f using the derivative is standard.[9] (Idea: break the number line up into regions on which f is "increasing" and "decreasing.") Graphing calculators may make this skill may go the way of the dodo, but the underlying idea is still important, and "$f' > 0$ implies f increasing" comes from the MVT.
   First we need some definitions.

Definition 9.3.3 *A function defined on (a, b) is <u>increasing</u> on (a, b) if $f(x_1) < f(x_2)$ for any x_1 and x_2 in (a, b) such that $x_1 < x_2$. It is <u>decreasing</u> on (a, b) if $f(x_1) > f(x_2)$ for any x_1 and x_2 in (a, b) such that $x_1 < x_2$.*

Draw some pictures to explore. One picture should show the appropriate inequality satisfied for some but not all pairs x_1 and x_2 in the interval to indicate why (graphically) the "*any x_1 and x_2*" is crucial.

9.41:

Aside
   Remark that our "increasing" is called "strictly increasing" in some texts; they reserve "increasing" for $f(x_2) \geq f(x_1)$ .... Some use "weakly increas-

---

[9]As is use of the second derivative, omitted here.

ing" or "nondecreasing" if equality is allowed. Check the definitions for safety.
End Aside

For hand graphing, "rising" and "falling" helps. Enter the derivative.

Proposition 9.3.4 *Let f be continuous on $[a, b]$ and differentiable on (a, b). If $f'(x) > 0$ for all x in (a, b) (written $f' > 0$ on (a, b)), then f is increasing on $[a, b]$, and if $f'(x) < 0$ for all x in (a, b), f is decreasing on $[a, b]$.*

The proof is an application of good quantifier habits and the MVT. To prove that f is increasing involves all pairs x_1 and x_2 in an interval, so let us pick an arbitrary x_1, x_2 so that $x_1 < x_2$. (Universal quantifier template; see Section 5.1.5.) Use $f'(x) > 0$ for all x in (a, b), apply the MVT and examine the positivity of various terms in its conclusion to finish.

9.42:

Since x_1 and x_2 were arbitrary, the result follows for all such pairs, which is as required by the definition. The second claim is similar.

We omit other MVT applications, such as the second derivative and "concavity." But besides graphing applications, this is important in algorism analysis. When you ask your calculator to find an x where $f(x) = 0$, your faith in its ability is touching and borne out by experience, but luckily for you there are people who worry about such things. The MVT gives some of the basis for analysis of success, and failure, of algorisms.

9.3.1   Exercises

9.43: The proof of Proposition 9.3.1 by contradiction isn't mathematically wrong, but avoidable proofs by contradiction are disliked (see Pólya's *How to Solve It* [5] for why). A direct proof is via the universal quantifier template. To prove is that, for every x_1 and x_2 in (a, b), $f(x_1) = f(x_2)$. Imitate the template in the proof of Proposition 9.3.4 and use the MVT as in our original proof of Proposition 9.3.1 for a better proof of that proposition.

9.44: Proposition 9.3.1 may be overgeneralized to falsehood by extending it on the basis of its intuitive statement. Consider f defined by

$$f(x) = \begin{cases} 3, & x < 0, \\ 5, & x > 0. \end{cases}$$

Show that f' is zero at each point of its domain, yet f is not constant. Moral: the "defined on an interval" hypothesis is important (cf. Section 4.3).

9.45: There is a simple, standard function that is increasing on $[-1, 1]$ although its derivative is *not* positive throughout. Find it. [Hints: "not positive" is not the same as "negative," and one x such that $f'(x) \not> 0$ suffices.]

9.46: Having reduced potential extrema to $f'(x) = 0$ or $f'(x)$ DNE, we'd like to decide if x is a max, a min, or neither. One aid is the First Derivative Test. Claim: if f is continuous at c, $f'(c) = 0$ or $f'(c)$ DNE, and there exists $\delta > 0$ so that f' is positive on $(c - \delta, c)$ and f' is negative on $(c, c + \delta)$, then c is a local maximum point. Picture? "Local minimum point" version? Prove using Proposition 9.3.4.

9.47: Via the (almost) uniqueness of f given f' we may prove that the natural logarithm, ln, has "log-like" properties. Let a be some positive constant and consider the two functions $\ln(ax)$ and $\ln x$. Compute their derivatives. So the two functions differ by a constant: $\ln(ax) = \ln x + C$. Insert $x = 1$. Find similar proofs for other standard properties of logs.

10
Other Limits

We present a sample of the various ways in which limit may appear in new situations, partly for perspective on the past and partly as a preview. Included is enough of the integral calculus to show you its differences from the usual limits, and a taste of sequences and functions of several variables. We begin with something tamer.

10.1   Limits Involving Infinity

You'll someday see "$\lim_{x \to \infty} f(x)$," which looks harmless until you notice the "∞" instead of a. Your temptation to treat this as just another limit, with ∞ viewed as a number, must be firmly squelched, because ∞ isn't a number. A definition of "limit at infinity" is lurking, and needed.

Definition 10.1.1 *Let f be a real-valued function defined on* **R**. *Then* $\lim_{x \to \infty} f(x) = L$ *if for every $\epsilon > 0$ there exists M so that*

$$|f(x) - L| < \epsilon, \qquad \text{for all } x > M.$$

Why not treat ∞ as just another a? In the ordinary definition, we need values of f close to L for all x near to a (on both sides). Here we require values of f be close to L for all x "large" (M the measure of "large"). The x for which you are responsible form a half line. Also, a appears in the usual definition, while ∞ appears nowhere here. Different concepts, period.

Time for a picture. Back in Section 1.2.1 you drew a certain picture (with horizontal and vertical strips) to capture informally $\lim_{x \to a} f(x) = L$.

Modify for $\lim_{x\to\infty} f(x) = L$.

10.1:

Find an example of f and L so that "after a certain point" all values of f are very close to L.

10.2:

Give several examples of functions that do not appear to have "limits at infinity." One should have values of f increasing apparently without bound, and leaving any proposed L far behind. Another sort ought to have the property that some values are actually taken on infinitely often, but none of them is suitable for the limit. Draw pictures with ϵ strips and M's showing failure, too.

10.3:

Your example for f with a limit might well have been $\frac{1}{x}$, $L = 0$. The graph is convincing. Try $g(x) = \frac{\sin x}{x}$ with your calculator to see a more complicated picture. Find, for some numerical values of ϵ, appropriate values for M.

10.4:

Important exercise: sketch out a program of results for $\lim_{x\to\infty}$ to prove by analogy with our $\lim_{x\to a}$ history. Look for your goals in the exercises to follow.

10.5:

10.1.1 *Exercises*

10.6: Formulate the definition of $\lim_{x\to-\infty} f(x)$. Find at least one function f so $\lim_{x\to\infty} f(x)$ exists but $\lim_{x\to-\infty} f(x)$ does not. [Hint: polynomials and rational functions won't do it.] We pursue the (minor) modifications for this theory no further.

10.7: Our insistence in Definition 10.1.1 that f is defined on all of $\mathbf{R}$ is unnecessary. To capture the behavior of, say, x^{-1} as x gets large, who cares that $1/0$ is undefined, since 0 is not "large" anyway? We want that "eventually" f is defined and has values close to something. So improve the definition: $\lim_{x \to \infty} f(x) = L$ if, for every $\epsilon > 0$, there exists M such that f is defined on (M, ∞) (more is fine) and $|f(x) - L| < \epsilon$ for all $x > M$. We use this definition henceforth. Find f with a limit at infinity under this definition, but not the old one, and g with a limit at infinity under either definition. Which definition is more restrictive?

10.8: Prove, using the Exercise 10.7 definition, that if f and g each have a limit at ∞, then $f + g$ does, and it is the right one. Prove the right thing for cf, where c is a constant. Remark: nothing can prevent you from using our proofs of the similar results for limit at a point as guidelines.

10.9: Prove that if $\lim_{x \to \infty} f(x) = 0$, and g is a function defined on some set (N, ∞) and bounded both above and below there, it follows that $\lim_{x \to \infty} f(x) \cdot g(x) = 0$.

10.10: Prove $f(x) = x^r$ (for $r < 0$) has a limit at infinity.

10.11: Repeat for $f(x) = e^{rx}$ (for $r < 0$).

10.12: Give a (concrete) example of f with a nonzero limit at infinity.

10.2   Sequences

The intuition for what a mathematician calls a sequence is a list of values, such as $1, 4, 9, 16, \ldots$.[1] There is a first entry in the list, a second, a third, and so on, and a feature of the sequences most often studied is that they are infinite in the sense that they "keep going" ("..."). To give a specific sequence, we either have to use the notation above (and trust you to find the fifth term by pattern), or adopt some notation. Usually s_n is the nth term in the sequence, so the sixth entry in the sequence above (36? Sure.) we make definite by $s_6 = 36$. Indeed, $s_n = n^2$ for all $n = 1, 2, \ldots$ is an unambiguous definition.

10.2.1   The Definition of Sequence

Since we want to prove things, a definition is crucial, and intuitive discussion about "lists" won't do. Key: I told you about the sequence by telling you (a formula for) values, exactly the way you get told about a function. A sequence is just that, a function, but one defined only on the positive integers, denoted **N**.

---

[1] Don't call this a series, since mathematicians reserve that word for something else and are unreasonably touchy about it, even with strangers and guests.

Definition 10.2.1 *A <u>sequence</u> s is a function from* **N** *into the real numbers. We often denote s(n) by s_n, and call s_n the nth term of the sequence.*

Just for practice, reconcile all this on the sequence s where $s_n = 1/n$.

10.13:

For a picture of a sequence, draw its graph, with the usual x–y coordinate system and points plotted, but plotted only for positive integer values of x (i.e., n). Draw one for $s_n = 1/n^2$. Then construct more examples, including ones with all $s_n > 0$, all $s_n < 0$, some of each, an alternating pattern, a constant sequence, and sequences appropriate to call "increasing," "decreasing," and neither. Pictures please!

10.14:

10.2.2   Sequence Limits

You should be hardened by now to defining new limits. Compare $s_n = \frac{1}{n}$ and $t_n = n^2$ for intuition as to what to capture in the first, and discard in the second.

10.15:

Remarks if you read Section 10.1

If you did, the work to come will seem completely predictable. We ask if anything interesting happens for the sequence when n gets large, as opposed to what happens to f when x gets large. Indeed, compare the pictures (same set of axes) for $f(x) = 1/x$ and $s_n = 1/n$.

10.16:

All becomes clear when you realize that s and f are both functions, just with different domains. If we find M to go with ϵ for f, surely all the n larger than M satisfy the inequality $\left|\frac{1}{n} - 0\right| < \epsilon$ since these n are among the $x > M$. It is easier to show that s has a limit than that f does, since s is just f with most of its domain thrown away.

You ought at this stage to be able to guess the formal definition of limit for a sequence. Do so (and then compare with what we actually use).

10.17:

Notice how the sequence and limit at infinity proofs are virtually identical.

End Remarks

The limit of a sequence should capture what happens "far out" in the list (like, all values close to some L if n is large). The only task is to see how to describe "far out" or "large enough."

Definition 10.2.2 *Let s be a sequence. We say $\lim_{n \to \infty} s_n = L$ if for every $\epsilon > 0$ there exists a positive integer N such that*

$$|s_n - L| < \epsilon, \quad \text{for all } n > N.$$

Comparison to $\lim_{x \to a} f(x)$ shows some striking similarities, and a striking difference. We require the values s_n (the output of the function s) to be "close" to L, and closeness is measured by $|\cdot|$ and ϵ. This part is essentially as before. But not all the output of s is likely to be close to L, so we may limit our responsibility for n's. For $\lim_{x \to a} f(x)$ we had a little region of size δ around a as our zone of responsibility; here, the set of all n greater than some N. Before: given ϵ, find $\delta > 0$. Here: given ϵ, find N. The scratchwork to find N will be familiar.

Example: guess $\lim_{n \to \infty} 1/n$; with $\epsilon = .1$, find a good N; show it works. Draw the picture of the sequence with L, ϵ, and N.

10.18:

Now consider the case $\epsilon = .07$, construct the requisite N, and show it works. Then show $\lim_{n \to \infty} 1/n = 0$ by finding N for an arbitrary $\epsilon > 0$.

10.19:

This example in hand, what's next? In the exercises we sample both parts of the usual approach, general limit theorems (e.g., the sum) and limits of classes of sequences (e.g., rational functions of n). For more results, see an analysis course, using a deeper understanding of the structure of $\mathbf{R}$.

Remark also that there are mathematical objects called "series." The series definition captures the idea of adding up an infinite (not just finite) number of terms. For example, if you patiently add up longer and longer pieces of

$$\frac{1}{2} + \frac{1}{4} + \frac{1}{8} + \frac{1}{16} + \frac{1}{32} + \cdots$$

it will become apparent that something seems to be happening.

10.20:

The study of series actually can be reduced to the study of sequences; one forms the sequence of "partial sums," which for the above series would be $1/2, 1/2+1/4, 1/2+1/4+1/8, \ldots$. We say the series converges to sum s if the sequence of partial sums converges to s. While the study of series is useful (see, e.g., Section 7.5.2), except for Exercise 10.27 we omit it.

10.2.3 Exercises

10.21: Prove that if sequences s and t each have a limit then $s+t$ does, the expected one. Prove that if s has a limit at infinity, then $c \cdot s$ does, where c is a constant. (Surely, $s + t$ is the sequence such that $(s + t)_n = s_n + t_n$, which is exactly termwise addition. You define $c \cdot s$ for yourself.)

10.22: Suppose $\lim_{n\to\infty} s_n = L$. Prove that if we alter the first $K = 100$ terms of s, the resulting sequence s' still has limit L. Repeat for general K.

10.23: Prove that any sequence s of the form $s_n = n^r$ for some $r < 0$ has a limit at infinity. Repeat for $s_n = e^{rn}$ for $r < 0$.

10.24: We explore the relationship between the $\lim_{x\to a} f(x)$ and how f behaves on sequences s_n approaching a. Suppose $\lim_{x\to a} f(x) = L$. Consider some sequence s with limit a as n approaches infinity. Consider a second sequence with values $f(s_1), f(s_2), \ldots$. Does this sequence have a limit? What limit? Careful: what if some s_n is a?

10.25: (Continued) Prove that if $\lim_{n\to\infty} s_n = a$ and no s_n equals a, and $\lim_{x\to a} f(x) = L$, then the sequence $f(s_n)$ has limit L. [Please read the Hint one chunk at a time, if you have to resort to it at all.]

10.26: (Continued) The previous result is unsurprising: to be given that $\lim_{x\to a} f(x) = L$ is to have information about how f acts on *all* points near a. Since any sequence s_n is just some of those points, f must be well behaved on these. More interesting, and surprising, is the following.

Proposition 10.2.3 *Suppose f is a function and L is a number so that for* all *sequences s with limit a (none of whose terms equals a), we have $\lim_{n\to\infty} f(s_n) = L$. Then $\lim_{x\to a} f(x) = L$.*

The proof is hard enough that we do most of the work, leaving you to fill in steps. First, we will prove the contrapositive, namely that if f does not have limit L at a then there is some sequence s approaching a (no s_n equals a) so that the sequence $f(s_1), f(s_2), \ldots$ does not have limit L. [Intuition: if f is bad in the functional limit sense, then there is a sequence out there that exhibits badness of f in the sequence sense.] We construct the bad sequence term by term.

Since the limit of f at a is not L, there is $\epsilon_* > 0$ so that for any $\delta > 0$, we can't get $|f(x) - L| < \epsilon_*$ for all x satisfying $0 < |x - a| < \delta$. Try $\delta = 1$; it fails, so there is some x satisfying $0 < |x - a| < 1$ but $|f(x) - L| \geq \epsilon_*$. Let s_1 (the first term in our sequence) be this x. Now try $\delta = 1/2$; it fails, so there is an x satisfying some inequality, but such that $f(x)$ does some other thing. What are the "some inequality" and "some other thing"? Let s_2 be this x. Continue in this way with $\delta = 1/3$, $\delta = 1/4$, ..., generating the sequence s, one term at a time. What inequalities do s_n and $f(s_n)$ satisfy?

Could $\lim_{n\to\infty} f(s_n) = L$? If so, for any $\epsilon > 0$, in particular our old friend ϵ_*, we could find N so that something happens. Trouble: *no* term of the sequence $f(s_1), f(s_2), \ldots$ satisfies the required inequality. So there is no good N, and $f(s_1), f(s_2), \ldots$ does not have limit L. Is s is a sequence, none of whose terms is a, and with limit a? Done (pictures, please).

Application: return to Exercise 1.56, and show the limit does not exist using this result.

10.27: Consider the series $(1 - 1/2) + (1/2 - 1/3) + (1/3 - 1/4) + \ldots$. Simplify term by term. Form the sequence of partial sums. To what does the series converge? Rearrange the terms above to see "why."

10.3   Functions of Several Variables

Calculus begins with functions of one input (from $\mathbf{R}$) and yielding one output (again in $\mathbf{R}$). Ordinary life is full of functions with more than one input and/or output. For example, the temperature function for your state is probably a function of two space variables (three, including altitude) and time: three inputs, one output (in $\mathbf{R}$). The position of a physical object is a function of one (time) input and three (space) outputs. Economic models have lots of both. We will write inputs and outputs as ordered pairs, triples, or n-length lists ("-tuples"; technically, vectors, although we won't use that language, except to call a function with a "-tuple" output a "vector-valued function") so the temperature function might be T(x, y, t) = T, and position $r(t) = (x, y, z)$. We let $\mathbf{R}^n$ denote the collection of all n-tuples. To the extent that we can graph things, we use the usual rectangular coordinate systems in two or three dimensions as appropriate (it would be nice to "see" in higher dimensions, but most people can't[2]).

An obvious next step is limits; we'll take the definition for functions of a single variable and see what the underlying idea is. Review the basic definition, and try to put it informally; the word "close" is a good one.

---

[2]There are occasional stories that high-powered mathematician X can "see" four dimensions; while such claims are interesting, they seem a bit hard to verify.

10.28:

So we need new measures of closeness. For one-variable input or output we will stick with our usual measure, absolute value ($|\cdot|$). But if the input is a pair (x, y), how will we measure how close this point is to (a, b)?

10.29:

There is a natural candidate.[3] Via the Pythagorean Theorem and co-ordinates, the distance between (a, b) and (x, y) is $\sqrt{(x-a)^2 + (y-b)^2}$. Points at exactly distance 1 from (a, b) lie on a circle of radius 1 about (a, b); points with distance less than 1 to (a, b) form the interior of that circle (a disk). Draw (many) pictures in the plane to illustrate these ideas.

10.30:

Generalization to $\mathbf{R}^n$ is algebraically easy (throw in more terms) if visually hard. In $\mathbf{R}^3$, the distance is the length of the line segment connecting the points (the diagonal of a rectangular solid with the points at opposite corners). We will stick to two or three dimensions in our examples.

We denote the -tuples as $a = (a_1, \ldots, a_n)$; the single symbol a is easy to read, but $(a_1, \ldots, a_n)$ reminds us it is a -tuple. (In $\mathbf{R}^2$, we may sometimes write (x, y) instead of (x_1, x_2) out of habit, and similarly (x, y, z) in $\mathbf{R}^3$.) If f is a function from $\mathbf{R}^n$ to $\mathbf{R}^m$, we can simply write $f(x)$. But remember that $f(x)$ is an m-tuple, so we may write it $(f(x)_1, f(x)_2, \ldots, f(x)_m)$, and remember also that x itself is an n-tuple.

Let d_n denote the distance function for -tuples of length n, so

$$d_n((a_1, \ldots, a_n), (b_1, \ldots, b_n)) = \sqrt{(a_1 - b_1)^2 + \ldots + (a_n - b_n)^2}.$$

Here's the definition, perhaps *too* general and abstract to understand at first.

Definition 10.3.1 *Let f be a function from $\mathbf{R}^n$ to $\mathbf{R}^m$. We say f has limit L at a, where $L = (L_1, \ldots, L_m) \in \mathbf{R}^m$ and $a = (a_1, \ldots, a_n) \in \mathbf{R}^n$, if, first, there is some $\theta > 0$ so that f is defined for each x such that*

---

[3]There are (useful) others. Measures of closeness (distance) are called "metrics," and there are several on $\mathbf{R}^2$, including the "taxicab" metric, in which the distance between a pair of points is the sum of the horizontal distance and the vertical distance, as if measuring the distance by how far a cab would drive on a rectangular grid of streets. You'll study such things in analysis or topology.

$d_n(x, a) < \theta$ *except possibly for a itself, and second, for every $\epsilon > 0$* *there exists $\delta > 0$ so that for all $x = (x_1, \ldots, x_n)$ in $\mathbf{R}^n$, if $d_n(x, a) =$* $\sqrt{(x_1 - a_1)^2 + \ldots + (x_n - a_n)^2} < \delta$ *then*

$$d_m(f(x), L) = \sqrt{(f(x)_1 - L_1)^2 + \ldots + (f(x)_m - L_m)^2} < \epsilon.$$

This definition requires some exploration. Start with seeing whether it gives the usual definition of limit in the case $n = 1$ and $m = 1$.

10.31:

Next is to write down the definition in two special cases, namely $n = 1$ and $m = 2$, and then $n = 2$ and $m = 1$. These are the simplest cases that are not just $\mathbf{R}^1$ to $\mathbf{R}^1$. But note that we ignore the difference between the number 3 and the 1-tuple (3).

10.32:

This section been mostly about functions of several variables and little about limits. Examples and graphing techniques are usually taught in third semester calculus, and are mostly omitted here. But the first exercises give a way to get helpful if less sophisticated pictures, and later exercises include limits in the standard multivariate calculus pictures.

10.3.1 Exercises

10.33: An ordinary function graph combines domain and range information in a single picture: we plot the ordered pair $(x, f(x))$. For functions of several variables we quickly run out of pictures we can draw, but there is a simple alternative, which we illustrate first with an ordinary function from $\mathbf{R}$ to $\mathbf{R}$. (This is exactly a domain–range picture, introduced in Section 1.3.) Take $f(x) = x^2$. Draw the domain of the function as a single vertical line (fair: the domain is $\mathbf{R}$). Draw the codomain (output space) as a single vertical line to the right of the first. We know $f(2) = 4$, and we show this by an arrow with tail at 2 in the domain space and with head at 4 in the codomain. Draw more relevant arrows; how does the picture show f is not injective? Not surjective (onto)?

Limits may be included. Not only is $f(2) = 4$, but $\lim_{x \to 2} f(x) = 4$. This is surely an ϵ–δ statement. Associated with some $\epsilon > 0$ there is an interval about 4 (range space, please) into which values of the function must fall. Draw it. We can find a $\delta > 0$, giving a punctured neighborhood $\mathcal{N}$ about 2 in the domain space so that for any x in $\mathcal{N}$, $f(x)$ is in the ϵ-neighborhood about 4. In the picture, any arrow with its tail in the δ-neighborhood must have its head in the ϵ-neighborhood. Draw such a δ-neighborhood. (A hollow dot at 2 reminds us that we don't have to worry about $f(2)$ for limits.) Note that the set of points "close to L" and the set of points "close to a" are nicely shown.

Draw the picture for a function without a limit: consider f defined by

$$f(x) = \begin{cases} 1, & x > 0, \\ -1, & x < 0, \\ 0, & x = 0. \end{cases}$$

This fails to have a limit at zero. Using an appropriate picture, show 0 is not the limit of f at zero. Using another, show $\lim_{x \to 0} f(x) \neq 1$.

10.34: To extend the picture to a function from $\mathbf{R}^2$ to $\mathbf{R}$, we must draw the domain space of the function on the left and the range on the right. The range space is easy: $\mathbf{R}$ again. The domain space is the usual set of ordered pairs, graphed in the usual x–y Cartesian coordinate system. We can draw arrows again, with tails in the domain, heads in the range; the tails are now based at an ordered pair. Try such a picture with $f(x, y) = x + y$; start with numerical values if necessary.

To include limits is to include the set of points "close to" the point in the domain, and around the proposed limit in the range. The set in the range, consists of z such that $|z - L| < \epsilon$ is the open interval: the target. The δ set in the domain, of (x, y)'s (we hope) close enough to (a, b) so $f(x, y)$ is in the ϵ-interval, is the disk of radius δ centered at (a, b). Draw the picture for $f(x, y) = x + y$, the point $(2, 3)$, and the limit L. With $\epsilon = 1$, give a safe δ-ball. Find a value of δ that does not work, and exhibit a point (x, y) showing it fails.

Again examine all of this with a function without a limit. Cheapest is to take a one-variable function and borrow it for this context. Namely, let f be defined by

$$f(x, y) = \begin{cases} 1, & x > 0, \\ -1, & x < 0, \\ 0, & x = 0. \end{cases}$$

(Note y is completely ignored.) Clearly this function has no limit at $(0, 2)$; show the difficulty via pictures. At what other points is there no limit?

10.35: For $f: \mathbf{R} \to \mathbf{R}^2$, we simply interchange the roles of the two spaces in the previous exercise. The range space is now pairs, and the set of points close to a given (a, b) is the disk centered at (a, b); the domain set is just $\mathbf{R}$. Repeat the steps from the exercises above, for a function with, and another without, a limit.

10.36: For $f: \mathbf{R}^2 \to \mathbf{R}$ it is possible to generalize the usual graph of a function and limit picture. To graph, we need two dimensions for the domain, and a third dimension for the range, and the graph is the plot of the ordered triples $(x, y, f(x, y))$ (sometimes put "plot $z = f(x, y)$"). For example, for $f(x, y) = x^2 + y^2$, you plot, among other points, $(0, 0, 0)$ and $(1, 2, 5)$. In fact, the graph of the function looks something like that below (a searchlight facing up the z axis).

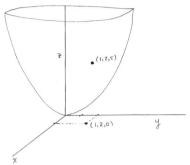

To introduce limits to the picture, consider the limit as $(x, y) \to (1, 2)$. Surely the limit is 5 (f ought to be continuous), and so associated with $L = 5$ and $\epsilon > 0$ there should be an interval on the range axis and some associated set in the picture as a whole (just like a little interval on the y axis and an associated horizontal strip for $f : \mathbf{R} \to \mathbf{R}$). Here the associated set is a horizontal slab of height 2ϵ centered at height 5. Points in this slab have z value within ϵ of 5, so this is the target.

Surely the whole graph of the function does not fit in this slab, so we must limit the points for which we are responsible to a small set S surrounding $(1, 2)$ in the domain. S is the set of points within δ of $(1, 2)$, a disk of radius δ centered at $(1, 2)$. (OK: we are not responsible for $f(1, 2)$, so it is a punctured disk of radius δ.) Goal: make the disk small enough so all (x, y) inside have the associated $f(x, y) = x^2 + y^2$ value land in the horizontal target slab. The domain set creates a vertical (punctured) cylinder in the graph, and in graphical terms we want the portion of the graph inside the cylinder also inside in the horizontal target slab. A familiar picture, if in higher dimensions! A graph, which anybody could imitate, is below.

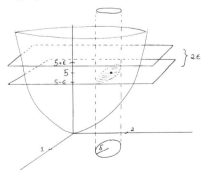

Draw similar pictures for the limit at $(0, 0)$. Consider $g(x, y) = 2y$; draw similar pictures for $\lim_{(x,y)\to(1,2)} g(x,y)$. Repeat for

$$h(x,y) = \begin{cases} 1, & x > 0, \\ -1, & x < 0, \\ 0, & x = 0, \end{cases}$$

at $(0,2)$ (no limit). Compare with Exercise 10.34.

10.37: The picture for $r\colon \mathbf{R} \to \mathbf{R}^2$ is quite different. We view r as giving a curve in the plane, as if viewing the input variable as time t and $r(t)$ as some point in the plane. My favorite image (leave me anonymous) is of a bug crawling along the curve and leaving "I was here at time t" markers every once in a while. So for $r(t) = (t, t^2)$ we get a curve in the plane (shape, the usual parabola) on which the bug passed through $(0,0)$ at $t = 0$, through $(2,4)$ at $t = 2$, and so on. Note that this picture is in the range space $\mathbf{R}^2$, and isn't a graph in the usual sense.

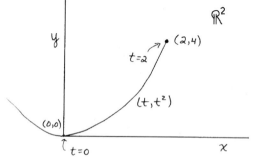

To include limit matters, consider $\lim_{t\to 2} r(t)$. The limit is $(2,4)$ (Continuity? Sure!). The collection of points close to it by ϵ is a disk centered at $(2,4)$ of radius ϵ, easy enough to draw. It's harder to include the domain, but remember that the t's of various points are marked along the curve (e.g., the point $(0,0)$ carries the label $t = 0$). So when we propose $\delta > 0$, we need all the points of the curve marked with labels between $2 - \delta$ and $2 + \delta$ (2 excluded) to lie in the ϵ-circle. The picture below (surely no proof) indicates success.

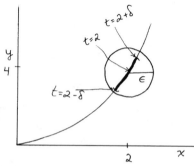

Draw similar pictures for the limit at $(0,0)$ and those for the bad function you constructed in Exercise 10.35.

10.3.2   *Limits for Functions from* **R** *to* **R**^{2}

To start, we do a ϵ–δ proof for concrete example, by hand as it were, to show you that the manipulations aren't too bad. Then we show that if $r(t) = (x(t), y(t))$, and x and y are good functions of t with respect to limits, then so is r, and conversely.

Consider $r(t) = (3t, t^2)$ and the limit, if any, at $t = 0$. You strongly suspect the limit is what? Next, write down what must be shown in this particular case (that is, take the general definition of limit, and insert our problem specifics).

10.38:

So let $\epsilon > 0$ be arbitrary. We seek a $\delta > 0$. We must arrange that $|t - 0| < \delta$ implies $\sqrt{(3t - 0)^2 + (t^2 - 0)^2} < \epsilon$. If no particular value of δ comes to mind (as why should it?), do the sensible exploratory thing: try $\delta = \epsilon$. (You either succeed, or you learn!)

10.39:

Hmmm: a mess, and not surely less than ϵ (since it is ϵ times something 3 or greater). But it has potential. We've gotten 2ϵ instead of ϵ in the past (see, e.g., Section 6.2). The trick is to modify δ to make the expression small enough to overcome the constant and yield ϵ at the end.

Inside the square root is $9 + \epsilon^2$, so a reasonable second try is $\delta = \epsilon/10$. Compute the quantity we hope ends up smaller than ϵ; does it?

10.40:

Including the (harmless) assumption $\epsilon < 1$, we get finally a satisfactory δ for ϵ, and are done.

Another approach is a "two-δ" view of the problem. If somehow $|3t - 0| < \epsilon/\sqrt{2}$, and also $|t^2 - 0| < \epsilon/\sqrt{2}$, then the expression $\sqrt{(3t - 0)^2 + (t^2 - 0)^2}$ would be less than ϵ. Check.

10.41:

This is sort of impartial; each coordinate function gets its fair share of the allowable error ϵ. So first we seek $\delta_1 > 0$ so that $|t - 0| < \delta_1$ implies $|3t - 0| < \epsilon/\sqrt{2}$. These are just one-variable limits, so work backwards. Then find δ_2 so $|t - 0| < \delta_2$ implies $|t^2 - 0| < \epsilon/\sqrt{2}$, assuming if you like that $\epsilon < 1$ or $\delta < 1$.

10.42:

We now have two δ's, and want a single one to handle both tasks, standard. Argue that this works, and get the inequality we need for the limit.

10.43:

This approach leads to our next result. Above, we found, by hand, δ_1 to go with $\epsilon/\sqrt{2}$ (making $|3t - 0| < \epsilon/\sqrt{2}$, i.e., handling the first coordinate function $3t$), and also δ_2 to go with $\epsilon/\sqrt{2}$ and (t^2). In retrospect, unsurprising: each of t and t^2 has a limit at $t = 0$, so of course such a δ_1 and δ_2 exist. Generalize: construct a function $r : \mathbf{R} \to \mathbf{R}^2$ from f and g, each with a limit at 0, by setting $r(t) = (f(t), g(t))$. For any $\epsilon > 0$, we could find for f a δ_1 for $\epsilon/\sqrt{2}$, and for g a δ_2 for $\epsilon/\sqrt{2}$. Check (easy) that with δ the minimum of δ_1 and δ_2 we have what is needed for a given ϵ to show that the limit for r exists, thus proving the (very useful) theorem below.

10.44:

Theorem 10.3.2 *Let f and g be functions with limits L_1 and L_2 at a, respectively. Then r defined by $r(t) = (f(t), g(t))$ for all t has limit (L_1, L_2) at a.*

The result (especially combined with the next one) yields some great corollaries.

We have proved that if x and y are good (as regards limits) then r built from them is good as regards limits. Other way? Yes.

Theorem 10.3.3 *Let f and g be functions from $\mathbf{R}$ to $\mathbf{R}$, and define a function $r : \mathbf{R} \to \mathbf{R}^2$ by $r(t) = (f(t), g(t))$ for all t. If r has limit $L = (L_1, L_2)$ at $t = a$, then f has limit L_1 at a and g has limit L_2 at a.*

To prove the result for f, find a δ to go with ϵ for f by using your ability to find δ to go with ϵ for r. Be simple.

10.45:

10.3.3 Exercises

10.46: First, define addition of points in $\mathbf{R}^2$ straightforwardly, by adding coordinatewise: $(x, y) + (a, b) = (x + a, y + b)$. Based on this, define the addition of functions from $\mathbf{R}$ to $\mathbf{R}^2$: if $r(t) = (x(t), y(t))$ and $s(t) = (a(t), b(t))$

then $(r+s)(t) = (x(t)+a(t), y(t)+b(t))$ for all t. Surely if r and s each have limits at t_0, then $r+s$ does. What limit? Proof?

The natural approach is to imitate past sum proofs, that is, an ϵ–δ approach, with δ the minimum of δ_1 (for r and $\epsilon/2$) and δ_2 for s and $\epsilon/2$. Try it; trouble awaits.

10.47: (That last exercise could have gone better.) Via our two theorems, we avoid the whole thing. Since r has a limit at t_0, so do x and y. Similarly, since s has a limit at t_0, (what?). Since x and a have limits at t_0, and as functions from **R** to **R**, their sum $x+a$ has (what?). Similarly, ... Now since $x+a$ and $y+b$ have limits at t_0, citing the first theorem .... Done.

10.48: Define continuity at a point for $r : \mathbf{R} \to \mathbf{R}^2$. Prove the continuity versions of Theorems 10.3.2 and 10.3.3.

10.49: Calculus for functions from **R** to $\mathbf{R}^2$ surely requires some sort of derivative, and there is an interesting choice of approaches. One is to say that since everything else happens coordinatewise (e.g., limits and continuity), the derivative must work that way too. This yields the definition $r'(t) \overset{\Delta}{=} (x'_1(t), x'_2(t), \ldots, x'_m(t))$ where $x'_i(t)$ is the usual one-variable derivative. Advantage: easy to compute; disadvantage: does it mean anything (geometrically, say)?

A second approach is to ignore the coordinates temporarily and return to the basic idea of derivative: take the difference of values of the function over shorter and shorter intervals, divide by the length of the interval, and take a limit. This gives

$$r'(t) \overset{\Delta}{=} \lim_{h \to 0} \frac{r(t+h) - r(t)}{h}.$$

(Note: this makes sense, since while you may not divide by a vector (-tuple), h is just in **R**, and with coordinatewise operations there is at least a chance for the limit to exist.) Try it on some numerical example, like $r(t) = (t, t^2)$ at $t = 2$. This definition has a good geometrical interpretation (if you are used to vectors, recall this produces a vector tangent to the curve mapped out by r). Disadvantage: is it easily computable?

Enter the theorems again. Show that these two potential definitions coincide, meaning that the limit for the second definition exists if and only if the limits implicit in the first definition (for the coordinate functions) exist, and the results are the same.

Note finally, and for future reference, that this "derivative" we have constructed is, since it is a -tuple, an object in the range space of the function.

10.3.4   Limits for Functions from $\mathbf{R}^2$ to **R**

You probably expect a rerun: an example by hand and some general result to do most of the work. We do the first, but the second step is much harder in this case and we just sketch what needs to be done.

Consider $f \colon \mathbf{R}^2 \to \mathbf{R}$ defined by $f(x,y) = x + y$, and suppose we want $\lim_{(x,y)\to(0,0)} f(x,y)$, surely 0. So we must show, by definition, that ...

10.50:

We need the usual scratchwork manipulation, and here is the crucial observation. Suppose $\sqrt{(x-0)^2 + (y-0)^2} < \delta$. Observe then that also $\sqrt{(x-0)^2} < \delta$, since this is no larger than the original square root.[4] Using $\sqrt{(x-0)^2} = |x - 0|$, we even have the usual absolute value control over x, namely $|x - 0| < \delta$, and similarly $|y - 0| < \delta$. To make $|x + y - 0| < \epsilon$, δ seems obvious. Show it works, carefully.

10.51:

Easy enough, but try $g(x,y) = xy$, say, at the point $(1,2)$.

10.52:

Ugh; there are occasional lucky coincidences of square roots and the squares: prove that $\lim_{(x,y)\to(1,2)} (x-1)^2 + (y-2)^2$ is what you think.

10.53:

Generally, though, the square roots and squares make a brute force approach unpleasant. Better is the usual battery of general "new limit from old" theorems about sums, products, and so on. State and prove the sum theorem.

10.54:

Results for polynomials follow from this, the product theorem (exercises), and showing that f defined by $f(x,y) = x$ (and the "y" version) has reasonable limits. Very useful for more complicated things is the following.

---

[4]This is independent (small modifications required) of the specific f or point $(0,0)$: control of (x,y) yields control over x and y individually.

Theorem 10.3.4 *Suppose f is a function from $\mathbf{R}^2$ to $\mathbf{R}$ with limit L at (a, b), and g is a function from $\mathbf{R}$ to $\mathbf{R}$ that is continuous at L. Then*

$$\lim_{(x,y)\to(a,b)} (g \circ f)(x, y) = g(L).$$

(See Section 6.5.3 for why g is assumed continuous, as opposed to merely having a limit.)

This lets old work give new results. To get $f(x, y) = \sin(x^2 y)$ continuous doesn't require an argument that considers both the function of two variables and the sine at the same time; we split f up as $g \circ h$, where $g(t) = \sin t$ and $h(x, y) = x^2 y$, and since the sine is good, we win if h is well behaved. Not having to consider the sine at the same time as handling square roots and squares by hand is a great benefit.

10.3.5   Exercises

10.55: Prove that if $f : \mathbf{R}^2 \to \mathbf{R}$ has limit L at (a, b), then there exists $\delta > 0$ so that for all (x, y), $(x, y) \neq (a, b)$ and $\sqrt{(x - a)^2 + (y - b)^2} < \delta$ implies $|f(x, y)| < |L| + 1$. Interpretation: a function with a limit at (a, b) is "bounded" (cannot grow arbitrarily large) near (a, b). Cf. Lemma 6.3.2 and Exercises 6.14 and 6.15.

10.56: Using Exercise 10.55, prove the product result about limits for functions from $\mathbf{R}^2$ to $\mathbf{R}$.

10.57: Prove Theorem 10.3.4. Draw the "domain–range" picture first.

10.58: Application: prove f defined by $f(x, y) = e^{xy} \sin(x^3 + xy)$ is continuous.

10.59: (Derivatives for functions from $\mathbf{R}^2$ to $\mathbf{R}$) (Note: much good geometric motivation omitted!) Given a function f, a point (a, b), and a nonzero vector (u, v),[5] we define the <u>directional</u> <u>derivative</u> of f at (a, b) in the direction (u, v) as

$$D_{(u,v)} f(a, b) = \lim_{t \to 0} \frac{f(a + tu, b + tv) - f(a, b)}{t \cdot \sqrt{u^2 + v^2}}.$$

(Assuming that (u, v) satisfies $\sqrt{u^2 + v^2} = 1$ cleans things up a little.)

This looks hard. But in spite of all this f, (a, b), and (u, v) stuff, since all these are fixed this is a *one* variable limit in t. Try $f(x, y) = x + 3y$ with $(a, b) = (0, 0)$, first with $(u, v) = (1, 0)$, then with $(u, v) = (0, 1)$, and finally with $(u, v) = (1, -1)$. Repeat with $g(x, y) = x^2 y$ at $(2, 3)$.

---

[5]If you are not comfortable with the language of vectors, think of this as simply another ordered pair in which at least one coordinate is not zero.

The <u>partial</u> <u>derivatives</u> (not yet "the derivative") of f at (a, b) are special directional derivatives. The partial derivative of f with respect to x is

$$\frac{\partial f}{\partial x} f(a, b) = D_{(1,0)} f(a, b)$$

and the yth partial derivative of f is

$$\frac{\partial f}{\partial y} f(a, b) = D_{(0,1)} f(a, b).$$

(What if there were three variables? We'd have x, y, and z partial derivatives, the last using the vector $(0, 0, 1)$ as the "(u, v, w)" vector (-tuple).)

To get "the derivative" of f we assemble into a -tuple some of these directional derivatives, namely, in order, $\dfrac{\partial f}{\partial x}$, $\dfrac{\partial f}{\partial y}$, and so on for higher dimensions. For $f \colon \mathbf{R}^2 \to \mathbf{R}$ we get the derivative, $\nabla f(a, b)$,

$$\nabla f(a, b) = (\frac{\partial f}{\partial x} f(a, b), \frac{\partial f}{\partial y} f(a, b)).$$

We have done nothing to make this appear even reasonable, let alone useful. Assemble $\nabla f(0, 0)$ and $\nabla g(2, 3)$ from your previous work anyway.

See a third-semester calculus book for real development. Observe (for our purposes) that in contrast to the case for $r \colon \mathbf{R} \to \mathbf{R}^2$, the derivative for $f \colon \mathbf{R}^2 \to \mathbf{R}$ is a vector in the domain, not in the range (see Exercise 10.49). Finding the (beautiful) unifying interpretation is part of an advanced multivariate calculus course.

10.4   The Integral

Your experience with the integral might be summarized as follows: it was defined in terms of limits of something called Riemann[6] sums, you may have computed exactly one such limit by hand, you proved a few general properties, someone said the integral exists for continuous functions, and then came the Fundamental Theorem of Calculus (argued or proved), which said you could compute integrals via antiderivatives. After this, integrals blurred with antiderivatives and you embarked upon some applications.

For a "careful definitions and proofs" text this would seem fertile territory. We'll do less than you expect because the proofs are rather deeper than the usual development makes clear (they've moved from introductory calculus into advanced calculus courses). Here we mostly point out where things are more complicated than they look, and in particular that the "limit" in "integral as limit of Riemann sums" is not at all the familiar kind.

---

[6] "Ree-mahn", please, not "Rye-man"; when you get your name on a theorem, you'll want it pronounced right, even after you're dead.

10.4.1   Difficulties with the Integral Definition

Recall the general setup: f defined on a closed interval $[a, b]$. Motivated by the approximation of areas by rectangles, it seemed of interest (or was declared of interest!) to partition $[a, b]$ into subintervals and form sums of terms "height of function times length of base." Thus, with the usual diagram below,

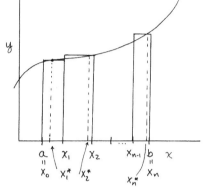

the sum is $f(x_1^*)(x_1 - x_0) + f(x_2^*)(x_2 - x_1) + \ldots + f(x_n^*)(x_n - x_{n-1})$, with the conventions $a = x_0 < x_1 < \ldots < x_{n-1} < x_n = b$ and that $x_j^* \in [x_{j-1}, x_j]$ for all j (that is, the point where we evaluate the function lies within the appropriate subinterval). To ease the notation, you defined $\Delta x_j = x_j - x_{j-1}$, the length of the jth subinterval, and wrote the "Riemann sum" as $\Sigma_{j=1}^n f(x_j^*)\Delta x_j$.

There were options and simplifications. Perhaps your tradition considered only partitions with equal-length subintervals, so that if there are to be n subintervals, each has length $\dfrac{b - a}{n}$ called simply Δx (a "regular" partition). There are various choices for the x_j^* at which to evaluate f, including

i) the left-hand endpoint of the subinterval;

ii) the right-hand endpoint of the subinterval;

iii) the point yielding the maximum value of f on the subinterval;

iv) the point yielding the minimum value of f on the subinterval;

v) the midpoint of the subinterval;

vi) an arbitrary x_j^*, so all we require is $x_j^* \in [x_{j-1}, x_j]$.

(Note that if f isn't assumed continuous, *iii)* and *iv)* may not be possible.)

The "definite integral of f from a to b" was then defined to be (something like) the limit of the Riemann sums as we force the lengths of the subintervals in the partition to become small. (Area pictures make this intuitively good.) But alarm bells should be going off now.

Let's take the simplest scheme. Using *regular* partitions (so there is only one partition for each n, the number of subintervals), *and one single* choice scheme for the x_j^* (any of *i)* $-v)$ above) things are actually OK. Question, though: what sort of limit are you taking?

10.60:

A sequence limit? Unexpected, and it raises all sorts of embarrassing questions. Does a different choice of the x_j^* (e.g., right-hand endpoints vs. left-hand endpoints) change the limit? Is there a limit? What about irregular partitions? And if the answer doesn't change, why make the definition specific for our choices of points and partitions, when it didn't really matter?

These difficulties are bad, but it's worse if we allow arbitrary partitions (even with just the left-hand endpoint approach, say). The difficulty is in the sort of limit. With regular partitions, there is a clear *order* to the choice of partitions, so we get a sequence of values. The partition with only one rectangle comes before that with two rectangles, and so on, and there is only *one* partition and result for each n. Fine, a sequence.

But take $[0,1]$ and allow arbitrary partitions. Consider the partition defined by the points $\{0,.5,.75,1\}$, and another partition using $\{0,.25,.5,1\}$. Which should come "before" the other in our passage to the limit? How do they compare to the partition defined by $\{0,1/3,2/3,1\}$?

10.61:

Moral: In past limits we've been unconscious of a crucial "ordering" of values. With sequences, surely the value for $n = 3$ came before that for $n = 5$, and the values for any n_1 and n_2 were in some comparable order: either n_1 was smaller and that one came first, or the other way around. Our limit definition used this, because we found N (a sort of fence) so all terms for $n > N$ did something (and we never worried about some n incomparable to N). Similarly, in our function limits, if $a = 2$, surely $x = 2.3$ is closer to $a = 2$ than $x = 1.6$. Again, our test for the limit involved setting a fence about a of size δ, and every x fell unambiguously either inside the fence or not; those inside concerned us, those not inside didn't.

One approach to the integral involves putting a sort of order on the set of partitions. Not all partitions are comparable, but some are, e.g., $\{0,.5,.75,1\}$ and $\{0,.25,.5,.75,1\}$. The second is called a <u>refinement</u> of the first; it has all the points of the first, as well as at least one more. We agree that the second is "greater" (farther out toward lots of very small subintervals) than the first. Fact: from any two (even incomparable) partitions we can construct a partition each can be compared to.

The notion needed is "common refinement." Given a pair of partitions, consider the partition defined by collecting all the points used for either. So for $\{0, .5, .75, 1\}$ and $\{0, .25, .5, 1\}$, the partition $\{0, .25, .5, .75, 1\}$ is a common refinement (as is anything using all these points and more). Note that $\{0, .25, .5, .75, 1\}$ is greater than $\{0, .5, .75, 1\}$, and is also greater than $\{0, .25, .5, 1\}$.

We are groping toward a "partial order," and with one a reasonable (harder) notion of limit follows. Given some L and $\epsilon > 0$, we seek a fence consisting of a partition P so all partitions greater than P have value within ϵ of L. We're stuck with some (lots of) partitions incomparable with P, but one can get an integral this way. (Note: all this is with a fixed choice scheme for the x_i^*.)

If we allow *both* arbitrary partitions *and* any method of selecting the x_j^*, all heck breaks loose. There is no good enough ordering on this mess to get limits (for a *single* partition, infinitely many x_i^* choices yield infinitely many values). After a few exercises, we turn to a new slant.

10.4.2 Exercises

10.62: Find a common refinement for $\{0, .25, .5, 1\}$ and $\{0, 1/3, 2/3, 1\}$. Given three partitions, what's a common refinement?

10.63: It turns out that monotone functions are good for integration however you define it. (Recall that a monotone function is one either all increasing or all decreasing on the interval in question.) For an increasing function, say, some of the choices for the x_j^* in *i) –vi)* actually coincide. Which? What about for decreasing functions?

Now compare the results of the maximum and minimum choices for a single regular partition. Illustrate with f defined by $f(x) = x^2$ on the interval $[2, 5]$ and the partition given by $\{2, 2.5, 3, 3.5, 4, 4.5, 5\}$ by computing the usual sum for the max choice, and then for the min choice. Don't simplify! What is the value of the "max sum" minus the "min sum?" Generalize the observation to an arbitrary regular partition for x^2, and then to an arbitrary increasing function f. If an integral exists at all, this gives upper and lower bounds for it.

10.64: Call a Riemann sum with maximum points chosen an <u>upper sum</u>. Suppose f, $[a, b]$, and P are given, and we form a refinement P' of P by adding a single point. Let U be the value of the upper Riemann sum for f and P, and U' for f and P'. Argue that $U' \leq U$ (that is, the refinement has lowered our upper sum approximation to the integral). Since any refinement of P can be gotten by inserting one point at each step, we can repeat the process to show that if U'' is the upper sum for f associated with any refinement of P, we have $U'' \leq U$.

Define <u>lower sum</u> and formulate the analogous result.

10.65: Let f, $[a, b]$, and a fixed partition P be given. Show that the upper sum for f and P (denoted $U(f, P)$) is at least as large as the lower sum $L(f, P)$ (assume f has maxes and mins). Again, this gives bounds for an integral.

10.4.3 Preparation: l.u.b. and g.l.b.

If limits aren't the right way to define the integral, what is? We use a definition that has nothing in particular to do with integrals, and just sketch it to hurry on to the integral, but it is vital for analysis (see an advanced calculus book).

Definition 10.4.1 *Let S be a nonempty subset of* **R**. *A* **least** **upper** **bound** *for S (denoted 'l.u.b.(S)' if it exists, and called elsewhere* **supremum** *and abbreviated "sup") is u such that for all s in S, $s \leq u$ (that is, u is an upper bound for S), and, for any $v < u$, v is not an upper bound for S.*

For some example sets S find the l.u.b., and find also S with no l.u.b..

10.66:

Could a set S have two distinct least upper bounds? (If so, the notation "l.u.b.(S)" is bad since such a symbol ought to point to a unique object.) No: if S has a least upper bound at all, l.u.b.(S) is unique (Exercise 10.71).

The cautious "if it exists" in the definition is troublesome. You have an example of a set that is not bounded above, so the question is not trivial. With a very careful development of the structure of **R**, one can prove that every nonempty subset of **R** that is bounded above has a least upper bound. But "**R**" is needed: consider an alternate universe in which the only numbers are the rational numbers. In that universe, consider $S = \{x : x^2 < 2\}$. Note that this set is bounded above (say, by 100). Does it have an l.u.b. in the universe in question?

10.67:

Filling in all "missing" l.u.b.'s ("holes") in the rationals does give the reals, but we leave this for an analysis course.

The l.u.b. (in **R** from now on!) has a useful property captured below.

Proposition 10.4.2 *Let S be a set with l.u.b.$(S) = u$. Then u is an upper bound for S, and for any $\epsilon > 0$, there is some s in S such that $s > u - \epsilon$.*

See the exercises for the proof (and more), but draw the number line picture now.

10.68:

Also note that the proposition lets us find a value in S "arbitrarily close" to l.u.b.(S) if we need one.

Investigate the relationship between "the l.u.b. of S" and "the maximum value of S." If S has a maximum M, is $M = $ l.u.b.(S)? If u is l.u.b.(S), is u a maximum?

10.69:

So for a set with no maximum, the l.u.b. is the best replacement.

Formulate the definition for the <u>greatest</u> <u>lower</u> <u>bound</u> (g.l.b.) of a set (elsewhere <u>infimum</u> and abbreviated "inf"), give a proposition indicating when there is a g.l.b., and formulate the analog of Proposition 10.4.2.

10.70:

10.4.4   Exercises

10.71: Prove that the l.u.b. of a set is unique if it exists. [Hint: suppose S has v and u each an l.u.b. of S. If $v < u$ use Proposition 10.4.2 for u the l.u.b., and contradict that claim that v is even an upper bound.]

10.72: Prove if $u = $ l.u.b.(S), then u has the properties in Proposition 10.4.2.

10.73: Suppose u is an upper bound for S, and for any $\epsilon > 0$, there exists some s in S such that $s > u - \epsilon$; prove u is an (hence "the") l.u.b. for S, so Proposition 10.4.2 is actually a characterization of the l.u.b.

10.74: Assume that any non-empty subset of $\mathbf{R}$ that is bounded above has a least upper bound; use this to prove that any nonempty subset of $\mathbf{R}$ bounded below has a greatest lower bound (the "g.l.b. existence proposition"). Idea (draw pictures): let S be a nonempty subset of $\mathbf{R}$ that is bounded below. Let T be the set $T = \{-s : s \in S\}$. Argue that T is bounded above, deduce it has an l.u.b., and get a great candidate for the g.l.b. of S. Prove it works.

10.4.5   Upper and Lower Sums

With l.u.b. and g.l.b., we may return to a "limit free" integral. We use l.u.b. and g.l.b. to replace the limits, because they are about sets of numbers and don't require any ordering of partitions.

Once and for all, we consider only f defined on $[a, b]$ and with the property that if $[c, d]$ is any subinterval of $[a, b]$, then f has a maximum and minimum value on $[c, d]$. Note that if f is continuous on $[a, b]$, this holds (Maximum Theorem), and if f is monotone on $[a, b]$, the appropriate endpoint of $[c, d]$ works.

Define $\mathcal{L}$ to be the collection of all lower sums $L(f, P)$ for the various partitions P of $[a, b]$, and $\mathcal{U}$ to be the collection of all upper sums $U(f, P)$. We want l.u.b.$(\mathcal{L})$ and l.u.b.$(\mathcal{U})$; do they exist? We need $\mathcal{L}$ bounded above and $\mathcal{U}$ bounded below. Argument 1: for each element of $\mathcal{L}$, say, $L(f, P)$ corresponding to some partition P, $L(f, P) \leq U(f, P)$ by Exercise 10.65, and therefore $\mathcal{L}$ is bounded above. Why isn't this right?

10.75:

The following lemma provides what we need and more: any lower sum is less than or equal to any upper sum. The proof (Exercise 10.85) uses a "common refinement."

Lemma 10.4.3 *Let f be as usual and P_1 and P_2 any partitions of $[a, b]$. Then*

$$L(f, P_1) \leq U(f, P_2).$$

So *any* upper sum is a satisfactory upper bound for $\mathcal{L}$, so $\mathcal{L}$ has an l.u.b. ($\mathcal{U}$ is similar).

Here are definitions crucial for the development of the Riemann integral.

Definition 10.4.4 *Let f be as usual. Define the upper integral of f from a to b, denoted $\overline{\int_a^b} f$, by*

$$\overline{\int_a^b} f = g.l.b.(\mathcal{U}).$$

Define the lower integral of f from a to b, denoted $\underline{\int_a^b} f$, by

$$\underline{\int_a^b} f = l.u.b.(\mathcal{L}).$$

Finally, we say that f is (Riemann) <u>integrable</u> on $[a, b]$ if $\overline{\int_a^b} f = \underline{\int_a^b} f$, and in this case we define the (Riemann) integral of f from a to b by

$$\int_a^b f = \overline{\int_a^b} f = \underline{\int_a^b} f.$$

Explore. In particular, find an informal and intuitive (if possibly imprecise) way of capturing the idea. Note: limits have completely disappeared.

10.76:

A function is not integrable if the values of the upper sums always stay very far away from the values of the lower sums, and so the g.l.b. of the upper sums can't be equal to the l.u.b. of the lower sums. Consider

$$f(x) = \begin{cases} 1, & x \text{ rational,} \\ 0, & x \text{ irrational.} \end{cases}$$

Compute the value of an arbitrary upper sum for f on $[0, 1]$. Compute the value of a lower sum for f on $[0, 1]$. Find $\mathcal{L}, \mathcal{U}, \overline{\int_0^1} f$, and $\underline{\int_0^1} f$.

10.77:

If f is integrable, the value of the integral may be trapped, as closely as we like, between $U(f, P)$ and $L(f, P)$ for some *single* partition P.

Proposition 10.4.5 *Let f be as usual and suppose f is integrable. Then for any $\epsilon > 0$ there is a partition P such that $|U(f, P) - L(f, P)| < \epsilon$ and*

$$L(f, P) \leq \int_a^b f \leq U(f, P).$$

Proof. Since $\int_a^b f = \text{l.u.b.}(\mathcal{L})$, cite Proposition 10.4.2 to show there is some ℓ in $\mathcal{L}$ such that $\ell > \text{l.u.b.}(\mathcal{L}) - \epsilon/2$. Let this element ℓ be associated with partition P_1, so $L(f, P_1) > \text{l.u.b.}(\mathcal{L}) - \epsilon/2$. Similarly, there is some partition P_2 such that $U(f, P_2) < \text{g.l.b.}(\mathcal{U}) + \epsilon/2$. Let P be a common refinement of P_1 and P_2; citing Exercise 10.64,

$$L(f, P) \geq L(f, P_1) > \text{l.u.b.}(\mathcal{L}) - \epsilon/2$$

and

$$U(f, P) \leq U(f, P_2) < \text{g.l.b.}(\mathcal{U}) + \epsilon/2.$$

Since f is integrable, $\int_a^b f = \text{l.u.b.}(\mathcal{L}) = \text{g.l.b.}(\mathcal{U})$, and the equations yield

(10.1) $$L(f, P) > \int_a^b f - \epsilon/2, \quad \text{and}$$

(10.2) $$U(f, P) < \int_a^b f + \epsilon/2.$$

From equations (10.1) and (10.2) it is easy to deduce the second claim of the proposition. As for the first, $L(f, P) \leq \text{l.u.b.}(\mathcal{L}) = \underline{\int_a^b} f = \int_a^b f =$

$\int_a^{\overline{b}} f \le U(f, P)$, where we have used f integrable and that l.u.b.$(\mathcal{L})$ is an upper bound for $\mathcal{L}$ and the analogous fact for $\mathcal{U}$. Done.

In fact, if we can do this sort of trapping, for any $\epsilon > 0$, then f is integrable, although we skip this here.

Sketch out a rough plan for the development of the integral.

10.78:

We hope your plan was a thorough one, since we won't carry it out. Our taste of limit free integrals turned into a full meal at least. But here is the theorem showing that limits of Riemann sums always give the right thing (proof omitted). Define first, for a partition P, $\|P\|$ to be the length of the longest subinterval in P (the <u>norm</u> of the partition P).

Theorem 10.4.6 *Let f be as usual, let f be integrable, and let $\{P_n\}_{n=1}^\infty$ be any sequence of partitions of $[a, b]$ such that $\lim_{n\to\infty} \|P_n\| = 0$. For each P_n, let $R(f, P_n)$ be a Riemann sum for f on $[a, b]$ obtained by making some choice of the x_j^* for the subintervals in P_n. Then*

$$\lim_{n\to\infty} R(f, P_n) = \int_a^b f.$$

Moral: *any* (reasonable) sequence of partitions (regular or not) and *any* choices of points (for example, any of $i) - vi)$ in Section 10.4.1) at which to evaluate the function yield values whose limit is the integral.

Making the norm of the partitions tend to zero is sensible. Draw some generic function on $[0, 1]$, and consider some sequence of partitions such that $[.5, 1]$ is always one of the subintervals, but we subdivide $[0, .5]$ more and more as n increases. The number of subintervals increases with n, but we aren't improving on $[.5, 1]$, and it is hard to expect the limit to be the integral.

10.79:

We omit most of the (extensive) theory of the integral, but do record a few results easy from previous work, and do a bit more in the exercises (including a proof of the Fundamental Theorem). Assume in what follows that continuous functions are integrable.

Proposition 10.4.7 *Suppose f is a function continuous on $[a, b]$ and with maximum M and minimum m on that interval. Then*

$$m \cdot (b - a) \le \int_a^b f \le M \cdot (b - a).$$

Alternatively, if u_m and u_M in $[a, b]$ are such that $f(u_m) = m$ and $f(u_M) = M$, we may write

$$f(u_m) \leq \frac{\int_a^b f}{b - a} \leq f(u_M).$$

Prove this, and draw suitable pictures.

10.80:

With this result, and the Intermediate Value Theorem for continuous functions, you can prove the Intermediate Value Theorem for Integrals.

Theorem 10.4.8 *Let f be a function continuous on $[a, b]$. Then there exists c in (a, b) such that*

$$\int_a^b f = f(c) \cdot (b - a).$$

(Interpretation: we could find a Riemann sum with a completely trivial partition yielding the value of the integral exactly, *if* only we knew how to find c, which of course we don't.) Draw the picture, prove the theorem.

10.81:

The Fundamental Theorem of Calculus is often viewed as merely a computational tool (often, the distinction between integrals and antiderivatives is blurred and integrals in their own right get lost). While vital as such a tool, Newton and Leibnitz (its independent English and German discoverers) get credit for the "invention of calculus" not because they invented derivatives or integrals (they didn't) but because they discovered this deep and unexpected relationship between derivatives and integrals. After all, the derivative deals with slopes of tangent lines, and the integral with areas under curves. Why should these be related? The great relationship is familiar, but a triumph of human intellect; don't be blasé.

Theorem 10.4.9 (Fundamental Theorem of Calculus) *Let f be continuous on the interval $[a, b]$. Then*

i) if F is defined on $[a, b]$ by $F(x) = \int_a^x f$, then $F'(x) = f(x)$ for all x in $[a, b]$, and

ii) if g is any antiderivative of f on $[a, b]$, then $\int_a^b f = g(b) - g(a)$.

Proof. See Exercise 10.86 for the proof of the first statement; given it, the proof of the second is easy. Since F and g are both antiderivatives of f, they differ by a constant: $F(x) = g(x) + C$ on $[a, b]$ (Corollary 9.3.2). We haven't defined $F(a) = \int_a^a f$, but the only sensible choice makes $F(a) = 0$. But then $\int_a^b f = F(b) = F(b) - F(a) = g(b) + C - (g(a) + C) = g(b) - g(a)$, as desired.

Remarks

First note that our development of the Riemann integral was more limited than necessary, just for ease of presentation. We assumed throughout that our functions had, on each subinterval of $[a, b]$, a maximum and minimum value; convenient, but not necessary. Things work if f is assumed merely bounded (above and below) on the interval $[a, b]$, and we replace maxes by l.u.b.'s and mins by g.l.b.'s, although we skip the details.

Another reason not to pursue the most general Riemann integral, is, alas, that the Riemann integral is not the best integral out there. There are many different definitions of "integral," and the Lebesgue integral is generally considered best. Thorough study of the Riemann integral, and great effort to get the most general Riemann integral, is rather like trying to learn about automobiles by making sure you have the most advanced Model T Ford. Unfair, perhaps, since the Riemann integral is used for lots of applications to this day. But for study of theory, it isn't the right thing to study in depth.

Two questions arise: first, "what is the Lebesgue integral?" See an introductory analysis course, but the idea is simple enough in general terms. Our development of the Riemann integral in terms of "area under a curve" approximated some general function by functions so easy that we know the area under them, and took a limit (well, all right, a g.l.b.) of such approximations. Our good functions are constant on subintervals, and the area under such a function is simply the height of the function times the length of the base. This gives the usual sort of rectangular approximation:

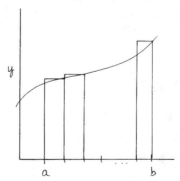

The Lebesgue integral uses "height of the function times the length of the base." However, the collection of suitable "bases" (subsets of $\mathbf{R}$ whose "length" we know) is vastly expanded by a study of what the length of a set should be. The "measure" of a set is a generalization of length to much more complicated sets (giving the usual answer for intervals). We then can try to integrate functions not well approximated by constant functions on intervals, but which can be well approximated by constant functions on these more general bases. The resulting collection of Lebesgue integrable functions is much larger than that of Riemann integrable functions.

The second question arising from deficiencies of the Riemann integral is "what makes a better integral?" The key is implicit in the last sentence of the preceding paragraph: we want an integral that integrates lots of functions. Of course, this quest can be too simpleminded. One possible definition of the integral makes the integral of every function, over every set, equal to zero. Well, fine. We want a definition, then, integrating lots of functions, but giving the "right" answers for simple functions (the Lebesgue integral does).

A Lebesgue integrable, but not Riemann integrable, function, is

$$f(x) = \begin{cases} 1, & x \text{ rational}, \\ 0, & x \text{ irrational}. \end{cases}$$

There are infinitely many points at which the function is 1, and infinitely many points at which it is 0, and the right value of the integral may not be obvious. But there are different sizes of infinity, and the infinity of points where $f = 1$ turns out to be a puny infinity compared to the set where $f = 0$. We've seen that this function is not Riemann integrable, but it is Lebesgue integrable. Indeed, let S be the subset of $[0, 1]$ on which the function has value 0. The measure ("length") of S is 1, and so the function is 0 on essentially all of the interval $[0, 1]$, and therefore has integral 0. See a text on "measure and integration" for details.

All this said, the Riemann integral is probably the more important; your physics, or economics, or engineering, may never require more. And the Riemann integral is better than you think; the function f (not continuous) defined by

$$f(x) = \begin{cases} 1, & x = .5, \\ 0, & x \neq .5 \end{cases}$$

is quite happily Riemann integrable on $[0, 1]$. Draw the picture, see what the upper and lower sums are going to look like, verify that it meets our assumptions, and find the value of the integral.

10.82:

10.4.6 Exercises

10.83: Intuitive from area pictures is that if $f(x) \geq g(x)$ on $[a, b]$, then $\int_a^b f \geq \int_a^b g$. Show first that for any partition P, $U(f, P) \geq U(g, P)$. Then show $\overline{\int_a^b} f \geq \overline{\int_a^b} g$. (This is a g.l.b. statement, so work at that level.)

Next, show that if both f and g are integrable, then $\int_a^b f \geq \int_a^b g$. Can you find an example of such functions f and g with equal integrals, even though $f(x) > g(x)$ for at least one x in the interval?

10.84: Let f and $[a, b]$ be given. Another standard fact, still intuitive from area, is that if P is a partition and P' is a second partition with all the same points of P, but more, $U(f, P') \leq U(f, P)$. Argue this (pictures help) if P' is gotten from P by merely adding one point. By repeatedly throwing in points, you get a chain of equalities giving the result for general P and P'. Formulate the result for lower sums.

10.85: We now seek $L(f, P_1) \leq U(f, P_2)$ for any P_1 and P_2. Here's the idea: find a common refinement P of P_1 and P_2. Use the previous exercise.

10.86: (Fundamental Theorem of Calculus) We use in the proof some results about integrals, such as

$$(10.3) \qquad \int_a^c f = \int_a^b f + \int_b^c f$$

and things gotten by subtraction from the above. The proofs (omitted) are neither extremely hard nor extremely interesting except as part of a thorough development of the integral (if then).

Recall we want that with F defined by $F(x) = \int_a^x f$ for $x \in [a, b]$, $F'(x) = f(x)$ for all x in $[a, b]$. This equation nicely provides a candidate for the limit, so we use difference quotients for F and an ϵ–δ approach. For notational reasons, we switch to the effort to show, for an arbitrary c in $[a, b]$ (universal template), that $F'(c) = f(c)$. So for every $\epsilon > 0$, we need $\delta > 0$ so $0 < |x - c| < \delta$ implies

$$(10.4) \qquad \left| \frac{F(x) - F(c)}{x - c} - f(c) \right| < \epsilon.$$

Well, $F(x) - F(c) = \int_a^x f - \int_a^c f = \int_c^x f$, by subtraction from (10.3). Assume temporarily $x > c$. Drag in the Mean Value Theorem for Integrals on the interval $[c, x]$ (watch notational traps). Out comes u in a certain interval, and by substituting we can turn inequality (10.4) into one involving only f, c, and u. It would be nice if a certain quantity were less than ϵ.

Intuitively, if u is close to c then something would be less than ϵ. What property of f is being used? Unfortunately, the location of u is not well specified: it is only known to be somewhere between c and x. So to force u close enough to c, we force x close enough to c, and we know there is some measure of "close enough."

The proof discovery over, write the proof. Include the (easy) changes for $x < c$.

Appendix A
Hints for Selected Exercises

Chapter 1
Section 1.1.1

1.1. Congratulations for actually doing something when you saw one of these – that's a great first step. It's better to put some solid work in before you consult these hints, so try some examples yourself before reading on.

Consider first the case in which $L = 0$ and $\epsilon = .1$. Test some points, such as 1, .5, and .05, and their negatives. Now consider $L = 5$, $\epsilon = .7$, and try various values of z. Are the relevant other values to try the negatives of the z you chose? What are the real analogous values?

Section 1.2.1

1.9. The picture looks a little like a square plaid.

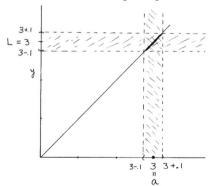

The picture in hand, plot on it some specific points on the graph of the function in the horizontal strip and the vertical strip and some others outside both. Since the limit is what we say it is, there are no points in the vertical strip not in the horizontal strip, and (very simple function) there are no points in the horizontal strip not in the vertical strip. Even if there were, this wouldn't discredit your claim that the limit is what you think.

Section 1.2.2

1.12. Draw the picture carefully (a large scale version, perhaps). Find some point on the graph of the function, inside the vertical strip, yet not in the horizontal strip. Note that in symbolic terms this is an x such that $0 < |x - a| < \delta$ but $|f(x) - L| \not< \epsilon$, or, in graphical language, the pair $(x, f(x))$ is in the vertical strip but not the horizontal.

Section 1.3.

1.15. You should be trying values of x close to 3 and evaluating f at them (say, $f(2.9) = 11.6$, and so on). You may have been simply plugging 3 into f. The answer is right, but the method has nothing to do with the notion of limit. Understand why the $f(2.9)$ idea is the right approach.

Section 1.3.1

1.21. Crucial question: how do choices of δ for this function compare to the choices for $4x$ as discussed in the text? Pictorially what is the difference between the functions, and does it make any difference for limit analysis at $a = 3$?

The most obvious wrong choice of limit is -2, which is $f(3)$. To show that the wrong choice of limit doesn't work, realize that for any reasonably small choice of ϵ, the $f(x)$ from some x (arising from a reasonable choice for δ) will be 12, and so can't also be close to -2. Indeed, how large an ϵ (roughly) would be needed to make some x close to 3 have $4x$ within ϵ of -2? (Answer: $\epsilon \approx 15$.)

1.23. Picture first. You were given a horizontal strip and found a vertical strip such that all points on the function graph inside the vertical strip fell inside the horizontal one. Then some very kind person decides to widen the horizontal strip. Isn't your old vertical strip fine? "Thanks!"

In algebraic terms, surely if $|f(x) - L| < \epsilon_{old}$, and $\epsilon_{old} < \epsilon_{new}$, then $|f(x) - L| < \epsilon_{new}$. That's for one x, but it's the same for all x in a set.

1.24. Your δ is more plausible if you notice that f is increasing (at on least relevant intervals). So if you find a $b < 2$ such that $b^2 > 2^2 - .1$, all x in the range $b < x < 2$ seem safe. So a candidate for your "left" δ is $2 - b$. Repeat to the right. Which is larger, your "left" δ or your "right" δ? Which therefore is the safe one when a single candidate for δ is needed (i.e., which makes you responsible for fewer x values when used as the δ in "both directions")? Draw the relevant pictures!

Moral: if ever again there are two candidates for δ, the smaller of the two is very likely to be the safe one. Hold this thought.

1.26. Graph the function and then use the TRACE feature (or whatever it is called) on your calculator. As you move away from a ($a = 2$ in this case) you get the coordinates of points on the curve, both the x and $f(x)$ values. When $f(x)$ gets too far from your limit ($L = e^2 \cong 7.389$), where "too far" is measured by ϵ, you have gone too far in x. Also, the graph (if trusted) indicates that there are no surprising bumps seeming to yield $f(x)$ values far from L for some x close to a.

Another approach is trial and error, by simply computing $f(x)$ at lots of points getting close to $a = 2$. This is actually what your calculator is doing when it graphs; doing it by hand makes obvious that we aren't even coming close to scrutinizing all points, so we get evidence, not proof.

1.29. A satisfactory δ can be found, but *not* by assuming that if some point b to the right of a has $f(b)$ close enough to L, then all points in the interval (a, b) will too. There are lots of points (infinitely many!) where $x \cdot \sin(1/x)$ is zero (the limit) but then the function wanders away again. The limit still exists, but not in a tame sort of way.

Section 1.4

1.32. Your library of functions ought to include polynomials and rational, trigonometric, exponential, and logarithmic functions. Also use piecewise defined functions such as

$$f(x) = \begin{cases} 3x, & x < 0, \\ 0, & x > 0. \end{cases}$$

Section 1.4.1

1.45. Begin with $\epsilon = 10$. Find a value of δ satisfactory with $L = 1$, and also a value of δ satisfactory for $L = -1$. The point is that the (correct) limit L must pass some test for *every* $\epsilon > 0$, and this shows that several potential L could pass the *single* test with $\epsilon = 10$.

Section 1.5

1.47. Graphical approach, graphing calculator. Graph and trace, and clearly values of the function are not settling down near zero (indeed, graphing for $x > 0$ suffices), but oscillate wildly. And even in small intervals around 0 (candidates for δ regions) the oscillation seems just as wild. Via a "ZOOM" feature on your calculator, or by regraphing with smaller range of x, this shows up vividly.

1.50. A graph via *Mathematica*©[6] is below. But even just numerically, we could get the string of points corresponding to $(1/(\pi/2), 1)$, $(1/(5\pi/2), 1)$, ... These alone would show that the limit can't be 0, since for $\epsilon = .1$ no choice of δ seems to exclude all of these from your responsibility.

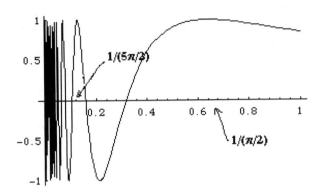

Section 1.5.1

1.56. The graph looks like two horizontal line segments, each full of holes. The holes are so small, but so many, that we can't really draw the graphs. But plot some points numerically, both for rational x and for irrational x.

If we suppose for the moment that $L = 1$, pick $\epsilon = .1$ and try some δ. Somewhere in the set of x such that $0 < |x - 0| < \delta$ there is an irrational number, whose function value is not anywhere close to 1. $L = 0$?

1.57. TRACE on your graphing calculator; try with numerical values of ϵ. Graph from $x = -.01$ to $x = 1$, $y = -1.2$ to $y = 1.2$; my calculator gives a low point at about $x = .222$, $y = -.217$, and thereafter the values of the function appear to be closer to zero than this. So if ϵ were .3, say, then $\delta = .222$ seems to limit my x values sufficiently, at least for $x > 0$.

For general ϵ the problem is harder, but note the apparent high and low points on the graph: they occur when x and y are roughly ...what? If so, associated with some $\epsilon > 0$, what is a safe δ? Argue analytically that your choice of δ is safe.

1.58. Graphing calculator; one can do a lot of useful work with only the graph for $x > 0$. Note that the graph of this function is always "lower" than for the previous one (at least for $0 < x < 1$, say). Given $\epsilon > 0$, and an associated δ satisfactory for the function of the previous exercise, how is δ is for ϵ and the present function?

Show algebraically $x^2 \cdot \sin(\frac{1}{x})$ is closer to 0 for $0 < x < 1$ than $x \cdot \sin(\frac{1}{x})$.

1.59. Plotting very many individual points, you get what appears to be two lines ($y = x$, $y = 0$). That's not a function graph, but in fact each "line" has holes in it just matching the points of the other. The jumping from "line" to "line" is not random; if x is rational we are on the upper one, and if x is irrational, on the lower.

A limit picture show that since each of the functions we are picking points from has limit 0 at $x = 0$, so does this much messier thing. Indeed, $|f(x) - 0| < \epsilon$ is completely trivial for x irrational. For rational x, check that a small δ will limit us to x such that $f(x) = x$ satisfies $|f(x) - 0| =$

$|x - 0| < \epsilon$. It will. Indeed, given some ϵ, what is a satisfactory δ?

1.62. Draw graphs for various values of b, and stop reading until you do.

Work also numerically. For values of x close to 4, but less, the values $f(x)$ are things like 3.9, 3.99, ... Since these are getting close to 4, the limit can't be anything but 4. Values of x close to 4, but greater, yield $f(x)$ like $-8.2 + b$, $-8.02 + b$, and so on. If we are to arrange that these are getting close to 4, we had better choose b sensibly.

Don't get the right answer for the wrong reason. If you said, "I need to make the two function values equal at $x = 4$, so we need $4 = -2 \cdot 4 + b$, ..., " you made the value $f(4)$ equal to the limit at 4. But function value is irrelevant to the limit, so this approach can't be right. (Yes, for friendly functions, $f(a)$ and $\lim_{x \to a} f(x)$ coincide, but think clearly.)

As for the δ–ϵ matters, find a δ_1 satisfactory for $-2x + 12$, and a δ_2 satisfactory for x. The minimum of these yields good x no matter which side of 4 x is on.

Chapter 2
Section 2.1
2.5. Classify failures of continuity at a: f might lack a value at a, lack a limit at a, lack both, or have a value and a limit but the value and the limit are not equal.

Note that the function in Exercise 1.59 is continuous at $x = 0$; elsewhere?

2.6. A graphing calculator is useful for the numerical examples, but the task of finding a satisfactory value of δ given some ϵ in general looks hard.

Section 2.2.1
2.19. For very many points a one can find an interval around a on which f is identically 0; fine, it is continuous there. For another class of points (see Exercise 1.60) there is a limit, and a function value, but these are unequal. (There's a limit since, except right at the point, no "unusual" points are very close by.)

There is one point without a limit, so f is not continuous there.

Section 2.3.1
2.22. Idea: you are responsible, in a left-hand limit, for only "half" the x values you would be for a standard limit. Originally you were responsible for every point in $(a - \delta, a + \delta)$ except a itself, now you get more exceptions: all of $[a, a + \delta)$. In graphical terms, you need only check whether points on the graph in the left-hand part of the vertical strip are in the horizontal strip.

So it's clearly easier to have a left-hand limit than a limit; show graphically why. Hint: armed with δ to accompany ϵ for a limit, what's a great candidate δ' for ϵ and a mere left-hand limit?

To find functions without left-hand limits, adapt a variety of the examples of functions without limits. (Of course, it is best to avoid functions failing to have a limit because there is a left-hand limit and a right-hand limit, but the values of these are unequal. This will *not* be a good source of functions without a left-hand limit.) Develop a full listing of functions without limits from the left (vertical asymptotes, a $\sin(1/x)$-style example, etc.).

Section 2.3.2

2.28. The approach via the definition and graphs is fine. Alternatively, realize continuity is purely a "local" definition; it has to do with the values of f at and near a. For this f, if we fix some $a > 0$, and look only at the graph in a little interval around a (be nearsighted!), everything looks exactly like the view of a nearsighted person at a but looking at g defined by $g(x) = 1$ for all x. This function is constant, hence continuous at a, and locally just like f near a, so f must be continuous at a too.

Section 2.3.3

2.33. Since the domain is not a closed interval, that special definition is irrelevant. Is f continuous at a rational number a? No, because the domain of definition doesn't include an open interval around a (surely there is an irrational in any such interval). So f is continuous at no points of its domain, hence not continuous.

Chapter 3

Section 3.1

3.2. Questions are not statements, nor are phrases without a verb. Further, sentences with a pronoun ("she") aren't statements, since until you know what the pronoun refers to you can't label true or false ("I am a red-head." True or False?). Later we develop something to turn such sentences into statements, but for the moment this is a reasonable rule.

Section 3.1.1

3.11. From a previous life you know both hypothesis and conclusion are true, so the implication is. But although this implication is true, the general form of the implication ("if f is continuous at a then f is differentiable at a") is not a safe one, since there are functions that are continuous at a point but not differentiable there. In ordinary language, "if–then" statements, if true, capture some kind of causality ("if it rains, I carry my umbrella"). But implications in mathematics require no connection between hypothesis and conclusion at all, and T or F are assigned to them solely on the basis of the T or F labels of their components, not component meaning or relationship. Look back at Exercise 3.8; what does the evenness of 24 have to do with the oddness of 37? Nothing.

3.12. Again the implication is true, exactly as outlined in the previous Hint; hypothesis true, conclusion true, check the grid, implication true.

But there is more going on here, as you probably remember. In general, if f is differentiable at a point then f is continuous there, and so this implication does capture a real connection between hypothesis and conclusion. Over and above of the rules of logic, this is the goal in mathematics: capture (true) connections between properties. The point is subtle but important.

Mathematical logic lets us combine statements any legal way we want, and tells us how to label them T or F. In practice, we expect to form *meaningful* statements while we play by these rules, not meaningless ones.

3.16. Little to say: these are the rules, and that's that. But the rules for "or" might be unexpected. In natural language we use "or" in two different ways. Any child understands that when a parent says "You may have a lolly pop or you may have some ice cream," both of the substatements are not going to be true simultaneously. This is the *exclusive* "or"; we assign value F to the compound statement if each of the sub-statements is T. This isn't the mathematical, *inclusive* "or", in which if both sub-statements are true the compound statement is labeled T. The natural language is sometimes "or both," but is equally often omitted entirely: "I'll take mathematics or English next semester (or both)." In mathematical logic we use the inclusive "or" throughout.

Section 3.2.1

3.23. This is one of those mathematical in-jokes; the quantification is obvious (the objects? positive integers). Here is a "proof" of this theorem:

Proof. Clearly 1 is an interesting number (for example, it is its own square root). If there were to exist any uninteresting positive integers, let w be the least of these. But to be the least uninteresting number is an interesting property for w, so w is interesting. Contradiction, so there are no uninteresting positive integers, so all positive integers are interesting.

Nice to know mathematicians can be as silly as anybody else.

Section 3.3.1

3.38. There are two ways to do this, one better but harder, one easier but less useful when proving. The easy way is just to translate the English:

$$(\forall \epsilon > 0)(\exists \delta > 0)(\ldots).$$

Problem: our definition of the universal quantifier was simply "for all x," not "for all x satisfying some condition" (same problem for "there exists"). It is standard to use the language above to capture this idea, but it is actually harder to prove things from, and it is unneeded.

The key idea for the universal is to insert an implication. In standard English, we take "for all $\epsilon > 0 \ldots$" and change it to "for all ϵ, if $\epsilon > 0$ then $\ldots$." For the existential quantifier, bundle the condition with the other thing(s) the object must do, using "and." Thus we might write "$\exists \delta (\delta > 0$ and $\ldots)$."

We may then write the above as "$\forall \epsilon (\epsilon > 0 \Rightarrow \exists \delta (\delta > 0$ and $\ldots))$." Or, if you prefer to use the notation $G(x)$ to mean $x > 0$, you may say "$\forall \epsilon (G(\epsilon) \Rightarrow \exists \delta (G(\delta)$ and $\ldots))$." (Either is correct, but note that on grounds of clarity the first is preferable; to encounter the notation "$G(x)$" is to have to remember what it means, while "$x > 0$" can be understood immediately.)

These latter forms will fit proof templates (to come) without translation.

Chapter 4

Section 4.1

4.2. The first broad division, excluding quantification, is the split of "first sentence forming the hypothesis" from "second sentence forming the conclusion." Write the conclusion with its quantifier next; with two conditions on x, write things as $\exists x(?$ and $?)$. Once you have written the universal quantification for f, remember to use the implication form to capture the (hint) three assumptions on f. Finally, remember that a and b are universally quantified: "a" closed interval $[a, b]$ is really "for any" closed interval $[a, b]$.

Section 4.1.1

4.16. Recall the classification: f may fail continuity at a by failing to have a limit (various subcategories here), by failing to be defined at a, or by having both limit and definition at a but lacking their equality. Each of these can make the conclusion of the theorem fail. Draw the pictures of such functions, and then pick the a and b as before.

Another approach is to return for a moment to the intuitive definitions of continuity involving "jump" and so on. Can these make the conclusion fail?

Section 4.1.4

4.21. Some interval like $[0, 3]$ works, since 2 is between $f(0) = 0$ and $f(3) = 9$. When we find x such that $f(x) = 2$, we have found a square root for 2. Nothing special about the square function, either. For two different roots, make the open intervals in the domain disjoint, and the values x_1 and x_2 so $x_1^2 = 2$ and $x_2^2 = 2$ will surely be different.

This won't guarantee just one (positive) root in any straightforward way, since the IVT guarantees at least one point, but does not say anything about only one point. Only one positive square root for 2 stems from other arguments.

Section 4.1.6

4.29. First, what does "approximate to two decimal places" mean? It can't mean "make sure that the first two decimal places of your answer match those of the correct value" (a reasonable guess) because most approximation schemes can't. Suppose that the actual answer is .75; we could be within one millionth of the correct value and still have first two decimal places .74 (having generated .7499999999999999999). So that's not what it means.

By convention, it means "with error less than half the value of the second decimal place," i.e., with error less than .005. The usual way to do this (at least in "interval reduction" methods, including the bisection algorism) is to obtain an interval of length less than .01 that is sure to contain the point, and then take its midpoint.

Question: how many iterations of the algorism will it take to achieve an interval of length less than .01?

Since each step of the algorism halves the length of the interval, we need to start with an interval of length 2 and halve it enough times to get one of length less than .01. Hmmm ... 10 halvings, a lot. What about one decimal place instead?

Section 4.2.3

4.42. This is a very important exercise, not from the point of view of mathematical content but for the development of "mathematical maturity." Put in some more time before you read any further.

We first separated the result into hypothesis and conclusion and made sure we understood each. We then produced some simple examples meeting the hypothesis and with hand-verifiable conclusion, and later some examples satisfying the hypothesis but for which the theorem really gave us something new (hand computation impossible). We checked, again by examples, whether violating parts of the hypotheses would allow us to evade the conclusion (this demonstrated the importance of the hypotheses). We finally constructed some examples in which the conclusion held even though the hypotheses were violated.

Section 4.2.4

4.56. Here's a useful approach: give yourself a function with a maximum, and then do (almost) anything you want in other places.

4.57. Via calculator, find the maximum and minimum points, so you know what closed interval is the range of the function on the set.

Section 5.1.1

5.2. The usual construction is to produce two circles of equal radius, each centered at one endpoint of the original segment. Their two points of intersection define a line, the right one. A certain minimum radius is needed.

Section 5.1.2

5.3. "Let $\delta = 1/2$" is the announcement of the candidate. The proof that δ is satisfactory is the next two sentences; note that "Thus $\delta = 1/2$ is as required" is purely a reminder of what we were doing, and not part of the proof that δ works but rather the marker for its completion.

Section 5.1.4

5.11. There appear to be two arrivals in the proof, and then the use of something derived from each of them as a candidate for something. Early in the proof "L" arrives, and "$-L$" is used as a candidate for the limit we want. Later in the proof we get a δ_* from the fact that f has a limit, and we use δ_* as a candidate for something we need for $-f$. What's going on?

Actually, nothing very exciting: there are two existence proofs going on. To show that $-f$ has a limit is to show a limit exists; as usual, we produce a candidate from somewhere. To show the candidate is satisfactory, we must show that for every $\epsilon > 0$ there exists $\delta > 0$, and this "nested" existence goes as usual: we get a candidate for δ and show it works.

Proofs within proofs (that is, the use of one proof template nested within another) can be confusing, but is absolutely standard.

Section 5.1.6

5.22. We will face proving $|x^2 - 16| < .1$, armed only with $0 < |x-4| < \delta$. We have $|x^2 - 16| = |x - 4| \cdot |x + 4|$, and the first factor is what we can make small. How large can the second be? How large if x is pretty close to 4? If we knew $|x+4|$ was surely less than 10, how might we choose δ? Can you do better with a better handle on $|x+4|$?

Section 5.2.2

5.31. One way to see the point of $\delta_* = \min(\epsilon_0/7, 1)$ is to see that δ_* must accomplish two things. One is to ensure $|x-3| < \epsilon/7$; the other is to ensure $|x + 3| < 7$. Assuming these are accomplished, showing $|x^2 - 9| < \epsilon$ is an easy multiplication. But argue in the proof that $|x-3| < \delta \le 1$ does ensure $|x+3| < 7$.

Section 5.2.3

5.33. Luckily no fancy bounding is needed here. You may use what you have used before, which is that if $0 < x < 1$, then $x^3 < x$ (put differently, if x is close to 0 but greater, then x^3 is even closer to 0). But if you have a write-up without $\delta < 1$ ensured, then you are probably in trouble, since $x^3 < x$ is not true in general.

Alternatively, if you take $\delta = \sqrt[3]{\epsilon}$, you don't have to worry about $\delta < 1$. Show this works, so we may view x^3 in some way other than "x times something we can control" because the cube function is so simple. Don't count on this technique away from zero.

5.34. The struggle will be to bound $x^2 + ax + a^2$ with $a = 2$, so we really have to cope with $x^2 + 2x + 4$. Suppose we choose $\delta < 1$ for sure, so x will be in the interval $(1, 3)$ (using $|x - 2| < \delta$). What is an upper bound for x^2 for these x? For $2x$ for these x? For $x^2 + 2x$? What about the "+4?"

Don't read this unless you have to, but $\delta_* = \min(\epsilon_0/19, 1)$ works. You do need to prove some lemma about how $|x^2 + 2x + 4| < 19$ if $|x - 2| < 1$.

5.37. The argument for $L = 0$ we have essentially done before.

For $L > 1$, use the universal template, so let L_0 be some arbitrary $L > 0$. We need $\epsilon > 0$ with no satisfactory δ; draw the picture, with L_0 included. Values of x with $f(x) = -1$ occur frequently, and these seem pretty far away from L_0 since $L_0 > 1$. This should hint to you that $\epsilon = 1$ won't work. Show it doesn't; read no further.

To show $L_0 > 1$ is not the limit, with $\epsilon = 1$, we must show that no $\delta > 0$ is satisfactory. Suppose one is proposed; as our intuitive arguments in Section 1.5 show, there is some x such that $0 < x < \delta$ but $f(x) = -1$, and so it is not true that $|f(x) - L_0| < \epsilon$. Remember to remark at the end that since $L_0 > 1$ was arbitrary, the result holds for all $L > 1$.

The argument for L less than -1 is similar; finish by disposing of L in $(0, 1)$ and L in $(-1, 0)$ using, of course, the universal template.

Chapter 6

6.5. To show that $\lim_{x \to b} g(x) = L$ we must show that

$$(\forall \epsilon > 0) \exists (\delta > 0)(\forall (x)(0 < |x - b| < \delta \Rightarrow |g(x) - L| < \epsilon)).$$

So let $\epsilon_0 > 0$ be arbitrary. Since $\lim_{x \to b} f(x) = L$, we know that $\forall (\epsilon > 0) \exists (\delta > 0)(\forall (x)(0 < |x - b| < \delta \Rightarrow |f(x) - L| < \epsilon))$. So in particular, for ϵ_0 there is such a δ, say, δ_*. We claim that δ_* is as required for g.

To show this, we must show that $\forall (x)(0 < |x - b| < \delta_* \Rightarrow |g(x) - L| < \epsilon_0)$. So let x_0 such that $0 < |x_0 - b| < \delta_*$ be arbitrary. Since $0 < |x_0 - b|$, $x_0 \neq b$, and so we have $f(x_0) = g(x_0)$. And by the choice of δ_*, since $0 < |x_0 - b| < \delta_*$, we have $|f(x_0) - L| < \epsilon_0$. Substituting $g(x_0)$ for $f(x_0)$, we have $|g(x_0) - L| < \epsilon_0$, as desired. Since x_0 was arbitrary, we have the result in general, which shows that δ_* is satisfactory for g and ϵ_0. Since $\epsilon_0 > 0$ was arbitrary, the result holds for all $\epsilon > 0$, and so we satisfy the definition to show that $\lim_{x \to b} g(x) = L$, as desired.

6.6. Before, we had only one point to "avoid," and that was handled automatically by $0 < |x - b|$. Here there are many to avoid, and the way to do so is to choose δ_* so small that $|x - b| < \delta_*$ ensures that we do. But we also have to choose δ_* so something else happens, so we seem to have to restrictions on δ_*. How do we cope? Stop reading and try again.

Let δ_1 be so that for all x, $0 < |x - b| < \delta_1$ implies $f(x) = g(x)$. For $\epsilon_0 > 0$, construct δ_2 using f has limit L at b. Try $\delta = \min(\delta_1, \delta_2)$.

Section 6.1.3

6.17. The key inequality is this: if we have some x such that $|f(x) - L| < L/2$, then $-L/2 < f(x) - L < L/2$, and by adding L to both sides we get $L/2 < f(x) < 3L/2$. Good trick.

6.18. Don't read this too soon, but it is useful to insert $\epsilon = \dfrac{\epsilon_0}{2/(L^2)}$ into the definition of the limit of f at L, and use δ a minimum again.

Section 6.2

6.21. Let f_0 and g_0 be arbitrary functions with limits L and M respectively at b. We show $\lim_{x \to b} f_0(x) + g_0(x) = L + M$. So let $\epsilon_0 > 0$ be arbitrary; inserting $\epsilon_0/2$ into the definition of the limit of f, we obtain δ_1 so $\forall (x)(0 < |x - b| < \delta_1 \Rightarrow |f_0(x) - L| < \epsilon_0/2)$. Similarly, we get δ_2 such that $\forall (x)(0 < |x - b| < \delta_2 \Rightarrow |g_0(x) - M| < \epsilon_0/2)$. We claim that $\delta = \min(\delta_1, \delta_2)$ works for ϵ_0 and $f + g$.

To show this, let x_0 such that $0 < |x_0 - b| < \delta$ be arbitrary. Then $|f_0(x_0) - L| < \epsilon_0/2$ and $|g_0(x_0) - M| < \epsilon_0/2$. So $|(f_0 + g_0)(x) - (L + M)| \le |f_0(x_0) - L| + |g_0(x_0) - M| < \epsilon_0/2 + \epsilon_0/2 = \epsilon_0$, as desired. Since x_0 was arbitrary, the result holds for all x, and δ is as needed. Since $\epsilon_0 > 0$ was arbitrary, there is a $\delta > 0$ for each such ϵ, as desired.

Section 6.3

6.26. Included for completeness is the proof that a function with a limit at b is bounded near b. No peeking until you've tried.

Proof. Let f have limit L at b. Using the definition with ϵ set to 1, there is a $\delta > 0$ such that for all x, $0 < |x - b| < \delta$ implies $|f(x) - L| < 1$. We claim that $K = \max(|L - 1|, |L + 1|)$ and the punctured neighborhood $N = \{x : b - \delta < x < b + \delta, x \neq b\}$ are what we require.

To prove this, we must show that for any x in that neighborhood we have $f(x) \leq K$, so let x_0 be arbitrary such that $b - \delta < x_0 < b + \delta$ and $x_0 \neq b$. Then $0 < |x_0 - b| < \delta$, so we may deduce that $|f(x_0) - L| < 1$. Thus $L - 1 < f(x_0) < L + 1$ via an easy manipulation of absolute values. Now since $|L - 1| \leq K$, $L - 1 \geq -|L - 1| \geq -K$. And since $L + 1 \leq |L + 1| \leq K$, surely $L + 1 \leq K$. Combining the inequalities, $-K < f(x_0) < K$, yielding $|f(x_0)| < K$ as desired. Since x_0 was arbitrary, we have the result for all x, and K is an upper bound for $f(x)$ on N. Thus there exists a punctured neighborhood on which f is bounded above, and since f and b were arbitrary, the result holds for all f with a limit at a point.

Section 7.4

7.18. Don't peek too soon, but here is an outline for the proof, where we have tried to make clear the separation between the "for every a" part of things and the induction part of things.

Pf. Let a be an arbitrary real number.

> We will show that x^n is continuous at a for all n by induction on n.
>
> First, we show that x^1 is continuous at a.
>
> **(your work here to show this.)**
>
> Second, we prove the induction step, so let n_0 be a fixed positive integer.
>
> We must show that x^{n_0} continuous at a implies x^{n_0+1} continuous at a
>
> So assume x^{n_0} is continuous at a.
>
> **(your work here to show x^{n_0+1} is continuous at a.)**
>
> Since we have shown this, we have the implication, and since n_0 was arbitrary, we have completed the induction step.
>
> Therefore, by the Induction Theorem, x^n is continuous at a for each positive integer n.

Since a was arbitrary, the continuity of these functions holds for all a in $\mathbf{R}$, and we are done. ∎

7.26. The discussion of the induction-based proof of the theorem that a sum of continuous functions of any length is continuous isn't quite complete.

The proof discovery in the text was fine at an intuitive level, but it contains a hole made explicit by stating the induction very carefully. We want to show that for any a in the reals, for any n a positive integer, and for any collection $f_1, \ldots, f_n$ of functions, the sum $f_1 + \ldots + f_n$ is continuous at a. We'll assume that a is arbitrary and never have to worry about it again. Fine. We are left with "$\forall n (P(n))$" to prove by induction. But what is $P(n)$?

Unfortunately, $P(n)$ is the statement "for all collections of functions (of length n), each continuous at a, the sum is continuous at a." Problem: $P(n)$ is *itself* a universally quantified statement (*all*) (new to us, since to date we have proved things by induction which were simple unquantified statements). So really at the induction step we are proving

$$(\forall f_1, \ldots, f_n (f_1, \ldots, f_n \text{ each contus at } a \quad \Rightarrow \quad f_1 + \ldots + f_n \text{ contus at } a))$$
$$\Rightarrow$$
$$(\forall f_1, \ldots, f_{n+1} (f_1, \ldots, f_{n+1} \text{ each contus at } a \quad \Rightarrow \quad f_1 + \ldots + f_{n+1} \text{ contus at } a)),$$

(where "contus" is a common shorthand for "continuous"). Even at the $n = 1$ step, we should have been proving something universally quantified. What? Take a moment and write it out, both in English and formally.

Having noted the difficulty, we leave the details of a correct proof as a challenge problem, or to be returned to after more experience with quantifier manipulations.

Section 7.4.1

7.29. The key to doing the induction is the careful definition of a polynomial of degree n or less: we will define a polynomial of degree less than or equal to n as a sum of monomials (in one variable x) whose powers of x are all less than or equal to n and such that each power less than or equal to n occurs exactly once as the power of a monomial in the sum. Observe then that a polynomial of degree less than or equal to n is a sum of exactly $n+1$ monomials. Note also that some or all of the coefficients of the monomials might be zero, so we are agreeing that what you usually think of as $4x^3 + 1$ shall be written $4x^3 + 0x^2 + 0x^1 + 1$.

Why not define "polynomial of degree exactly n" (i.e., guaranteed non-zero coefficient of x^n)? The induction turns out to be trickier; continuity of $x^3 + 5x + 7$ should rest on x^3 continuous and $5x + 7$ continuous because $5x+7$ is of degree two (assumed continuous by induction hypothesis) ... but it isn't exactly of degree two. This can be made to work, but "degree less than or equal to n" is cleaner.

7.30. The limit of a sum of length 1 is easy; the limit of a sum of length $n + 1$ can be written as the limit of a "(sum of length n) + one term"; you know something about each of these.

Chapter 8
Section 8.1

8.6. The crucial string of equalities follows; provide a reason for each.

$$\lim_{x \to a} \frac{f(x) \cdot g(x) - f(a) \cdot g(a)}{x - a} =$$
$$= \lim_{x \to a} \frac{f(x) \cdot g(x) - f(x) \cdot g(a) + f(x) \cdot g(a) - f(a) \cdot g(a)}{x - a}$$
$$= \lim_{x \to a} \frac{f(x) \cdot g(x) - f(x) \cdot g(a)}{x - a} + \lim_{x \to a} \frac{f(x) \cdot g(a) - f(a) \cdot g(a)}{x - a}$$
$$= \lim_{x \to a} f(x) \cdot \frac{g(x) - g(a)}{x - a} + \lim_{x \to a} \frac{f(x) - f(a)}{x - a} \cdot g(a)$$
$$= \lim_{x \to a} f(x) \cdot \lim_{x \to a} \frac{g(x) - g(a)}{x - a} + \lim_{x \to a} \frac{f(x) - f(a)}{x - a} \cdot \lim_{x \to a} g(a)$$
$$= f(a) \cdot g'(a) + f'(a) \cdot g(a).$$

In composing your reasons, note that our rules about limits (of a sum, say) have the flavor "if limit A and limit B exist, then the limit of A + B exists, and is the limit of A plus the limit of B." So for many of your steps, you must include something like "we will show shortly that these limits actually exist." This is all shorthand for what we really ought to do, which is note that $\lim_{x \to a} \frac{f(x) - f(a)}{x - a}$ exists, and also $\lim_{x \to a} g(a)$ exists (just the limit of a constant, since a is fixed). Therefore, $\lim_{x \to a} \frac{f(x) - f(a)}{x - a} \cdot g(a)$ exists, and is equal to $\lim_{x \to a} \frac{f(x) - f(a)}{x - a} \cdot \lim_{x \to a} g(a)$. Repeat as needed. *Then* we could claim that the limit of the sum exists, and was equal to the thing we wanted to start with. In practice, as long as everybody in the room knows that the limits are going to exist in the end, the proof isn't written this way.

Also note that f, g, and a should be "arbitrary" (universal template).

Section 8.1.1

8.10. After you have formed the difference quotient for $\frac{1}{f}$, find a common denominator and then pull out the limit that is the derivative of f. Question: must you actually use the continuity of f at the point a? What hypotheses on f are required? Spend some time here, for there is a subtle point.

You almost certainly missed a hypothesis. Obviously $f(a) \neq 0$ is required. Much less obvious is that f being zero other places could be a problem, but it is. Now Definition 8.1.1 shows that our definition assumes that the function in question exists in an open interval surrounding a. Since f is assumed to have a derivative at a, it surely does, but unfortunately the function in question is $\frac{1}{f}$, and we have seen before a way in which f might exist throughout an interval, but $\frac{1}{f}$ might not. Right. f could be zero. Of course, f isn't zero at a, but could it be zero very often around a (recall $\sin \frac{1}{x}$ near 0).

See Exercise 6.16 and following for a discussion (facing a similar difficulty) showing that since f is not zero at a, and since f is continuous at a

since it is differentiable at a (our working assumption) then f is not zero in some open interval about a. So no additional assumption is required to keep f from being zero "around" a.

Section 8.4

8.24. A chain of equalities is as follows:

$$
\begin{aligned}
(f_1 + \ldots f_n)'(a) &= \lim_{x \to a} \frac{(f_1 + \ldots f_n)(x) - (f_1 + \ldots f_n)(a)}{(x - a)} \\
&= \lim_{x \to a} \frac{(f_1(x) - f_1(a)) + \ldots + (f_n(x) - f_n(a))}{(x - a)} \\
&= \lim_{x \to a} \frac{(f_1(x) - f_1(a))}{(x - a)} + \ldots + \lim_{x \to a} \frac{(f_n(x) - f_n(a))}{(x - a)} \\
&= f_1'(a) + \ldots + f_n'(a).
\end{aligned}
$$

Justify each step; quantifier templates are also needed for the proof.

Section 8.5

8.30. Observe that

$$
\begin{aligned}
\lim_{h \to 0} \frac{\sin(a + h) - \sin a}{h} &= \lim_{h \to 0} \frac{(\sin a \cdot \cos h + \cos a \cdot \sin h) - \sin a}{h} \\
&= \lim_{h \to 0} \left(\frac{\sin a \cdot (\cos h - 1)}{h} + \frac{\cos a \cdot \sin h}{h} \right) \\
&= \lim_{h \to 0} \frac{\sin a \cdot (\cos h - 1)}{h} + \lim_{h \to 0} \frac{\cos a \cdot \sin h}{h} \\
&= \sin a \cdot \lim_{h \to 0} \frac{\cdot(\cos h - 1)}{h} + \cos a \cdot \lim_{h \to 0} \frac{\sin h}{h},
\end{aligned}
$$

assuming we can later show certain limits exist. So the derivative of the sine at a comes down to evaluating these two limits, and we'll get the derivative of the cosine (and the rest) from this information.

Section 8.5.2

8.36. Let's rule out the absolutely horrible thing immediately. A way to get the right answer by doing the wrong thing is to "cancel" as follows: cancel the sines, leaving behind the $\frac{3x}{5x}$, and then continue, resulting in $3/5$ (coincidentally correct). This is silly, but surprisingly common among people who haven't really grasped that the sine is a function, and so it is the "sine of x" as opposed to the "sine times x." One can no more cancel sines like this than one can cancel the square root signs in $\frac{\sqrt{3x}}{\sqrt{x}}$ to eventually yield 3.

What's a real solution? The solution usually given is as follows:

$$
\lim_{x \to 0} \frac{\sin 3x}{5x} = \lim_{x \to 0} \left(\frac{\sin 3x}{3x} \cdot \frac{3}{5} \right)
$$

$$= \lim_{x\to 0} \frac{\sin 3x}{3x} \cdot \lim_{x\to 0} \frac{3}{5}$$
$$= \frac{3}{5} \cdot \lim_{x\to 0} \frac{\sin 3x}{3x}$$
$$= \frac{3}{5} \cdot 1.$$

Of course this is all fine except possibly for the last equality, which is usually justified something like this: "Since $3x$ approaches 0 as x approaches 0, we can cite equation (8.7) with $\theta = 3x$." Oh, really?

How are we going to justify this? It is really some statement, in a brilliantly confusing disguise, about the limit of a composite function. The general claim is something like "if $\lim_{x\to a} g(x) = L$ and $\lim_{x\to b} h(x) = a$ then $\lim_{x\to b} g(h(x)) = L$. (Here $h(x) = 3x$ and $g(x) = \frac{\sin x}{x}$.) Although false as stated, it can be repaired; see Section 6.5.3 for a discussion. Check to see that armed with this result the above argument goes through. (The point is not that this is a crucial result, but that the usual justification is too offhand.)

Chapter 9

9.2. Without the "greater than *or equal to*" clause the definition is doomed, since otherwise we would require $f(a) > f(a)$. See why? Produce an example of a function in which the possibility of equality really occurs (start with a picture). With an explicit function, you may have a good function to show you that a point could be both a local maximum point and a local minimum point for a (very special) function.

Section 9.1.2

9.8. Any picture I can draw looks something like the following:

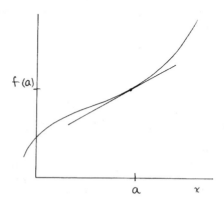

Apparently, points to the right of a have function values larger than $f(a)$. The fact that "local maximum point" only requires a to be king of the mountain for a very small region doesn't help, since even very close to a (on the right) are larger function values.

Section 9.1.3

9.15. Easy to miss in the formal write-up is showing your proposed punctured neighborhood actually works. (Pictures and intuition make it convincing, but that isn't a proof.) Recall from the definition that this involves showing that for all x satisfying $|x-a| < \delta$ we have $f(x) < L/2$ (remember, we want the punctured neighborhood to do something for f, not for $-f$). This, since it is about *all* x, requires the universal proof template from Section 5.1.5. Start by picking an *arbitrary* x satisfying $|x - a| < \delta$, and show $f(x) < L/2$. If we succeed, we get the result for all x, as needed.

Here's how it goes. Let x satisfying $|x-a| < \delta$ be arbitrary. By our choice of δ, we have that $-f(x) > -L/2$ (recall that δ was chosen to do something for $-f$). Multiply this inequality by -1: we get $f(x) = (-1) \cdot -f(x) < (-1)\cdot(-L/2) = L/2$, as required. Since x satisfying $|x-a| < \delta$ was arbitrary, we have the result for all such x by the universal proof form. Thus δ defines a satisfactory punctured neighborhood about a for f, as desired.

Section 9.2.4

9.36. Note that some extra work is required, since the simple approach of two applications of the MVT does not work: how are you sure that the points "c" for f and g are the same? What is needed is to produce a new, single function, so that a point for it will turn out to be a suitable (simultaneous) point for f and g.

9.37. We can indeed write the left-hand side as $\dfrac{\dfrac{f(b) - f(a)}{b - a}}{\dfrac{g(b) - g(a)}{b - a}}$. It certainly seems that this ought to be equal to $\dfrac{f'(c)}{g'(c)}$. But there's an error. Yes, there is some c so that $\dfrac{f(b) - f(a)}{b - a} = f'(c)$. And there is some z so that $\dfrac{g(b) - g(a)}{b - a} = g'(z)$. Unfortunately, there's no reason to believe that z and c are the same.

Indeed, consider $\sin x$ and $\sin 2x$ on $[0, \pi]$. The above attempt at an argument hopes for a single point where each has derivative zero. Via a picture, show this fails, and conclude that Cauchy's Formula cannot be proved by this approach.

Chapter 10
Section 10.1

10.1. Surely the horizontal strip is the same as in the usual definition, since it is associated with L and f and everything is the same there. The vertical strip is now a pretty impressive one, since it lies over the half line $\{x : x > M\}$. The intersection of the two strips is, as before, where the function must fall. Note finally that there is no "excluded" point in this definition; whereas before we didn't need to worry about $f(a)$ in this case

$f(\infty)$ isn't even defined.

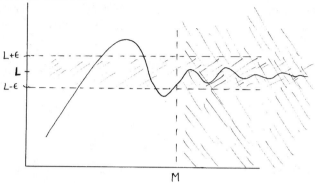

Section 10.1.1

10.11. (The manipulations for this problem are rather like those for the previous one.) We have some $\epsilon > 0$; we seek some M so that $x > M$ implies $|e^{rx} - 0| < \epsilon$. This becomes $e^{rx} < \epsilon$. Take the natural log of both sides to yield $rx < \log \epsilon$. Divide both sides by r, taking care with the sign of r.

This looks like a good candidate for M, but why is the expression on the right hand side positive? It won't be, unless $\log \epsilon$ is positive, i.e., ϵ is less than 1. Suppose it is; check that the candidate for M works.

What if $\epsilon \geq 1$? You avoid the problem: find an M to go with some $\epsilon < 1$, say, $\epsilon = .5$ just to be specific. Check that your M will work for the original ϵ. So only "small" ϵ are the problem, and it is standard to worry only about $\epsilon < 1$ if technically convenient.

Section 10.2.1

10.13. Viewing a function as a table of values, we would get:

n	$s(n)$ or s_n
1	1/1
2	1/2
3	1/3
4	1/4

(and so on).

10.16. The picture with both functions drawn on the same axis looks rather like a string of beads; the graph of $y = f(x)$ looks like the string, and if we enlarge the points on the graph for s, they look like the beads. Viewing the picture, for which (f or s) do you think it easier that there will be a limit as the input variable (x or n respectively) becomes large?

Section 10.2.3

10.24. Intuition says something like this: as n gets large, s_n gets close to a, and for values of x close to a we have $f(x)$ close to L, so $f(s_n)$ ought to be close to L. So we expect that $\lim_{n \to \infty} f(s_n) = L$.

Now we have to try to poke holes in this argument. One technical problem is that one sequence approaching a is the sequence $a, a, a, \ldots$. But just because f has a limit at a, that doesn't mean we can say anything about $f(a)$; it may be undefined, and if so, our effort to evaluate f at terms of the sequence above will fail. There's the same problem if even one term of the sequence is a. To fix this is to ask, instead of our original question, "What happens to the sequences of values of f at sequences approaching a, but such that no s_n is a?"

10.25. The first thing is to state your goal clearly. Given some $\epsilon > 0$, you have to find some N so that for all $n > N$ you have $|f(s_n) - L| < \epsilon$.

To aid you in performing the task above, you have the following powers: given some $\gamma > 0$, you may find some $\theta > 0$ so that if $0 < |x - a| < \theta$ then $|f(x) - L| < \gamma$. Also, given $\alpha > 0$, there exists M so that for all n such that $n > M$ one has $|s_n - a| < \alpha$. This forest of different Greek letters is to stress that you have, in some sense, complete flexibility to insert things into these limit definitions. Some insertions are unhelpful or lead to nonsense; explore.

The choices are to insert the ϵ into the definition of function limit, or into the definition of sequence limit. If you try the sequence limit (that is, insert ϵ for α), you have M so that if $n > M$ then $|s_n - a| < \epsilon$. This doesn't seem to be helpful, since it has nothing to do with f. Further, there is no way to insert M into anything to do with f. So try the other way.

If you insert the ϵ for γ, you get $\theta > 0$ so that if $0 < |x - a| < \theta$ then $|f(x) - L| < \epsilon$. You may argue that this has nothing to do with the sequence, and that's true. But θ can be inserted into the sequence limit (inserted for α, and out pops an M). Conceivably M is a suitable candidate for N. Show that it works.

Here's a picture.

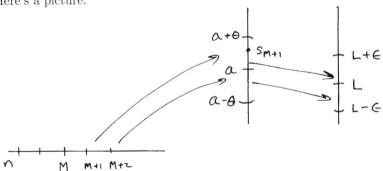

Remember that showing M works for N requires the universal quantifier template for n, so start by assuming that $n > M$. What can you deduce about s_n? Therefore about $f(s_n)$? Where's the observation and justification that $s_n \neq a$?

Section 10.3

10.32. There's some standard notation in these cases because the length of the -tuples is so small. First, if $f : \mathbf{R}^2 \to \mathbf{R}^1 = \mathbf{R}$, then we write $f(x, y)$ for the real number output of f. Such a function might be, for example, $f(x, y) = x + y$. Create a few more examples; one of them should be a function that completely ignores one of the inputs. Another should completely ignore both of the inputs (its name?). A final example of such a function might be the temperature on a flat plate.

For a function $r : \mathbf{R} \to \mathbf{R}^2$ we take a single input (often called t) and produce a pair of outputs. This is often written as $r(t) = (r_1(t), r_2(t))$. The idea is that the x coordinate (the first coordinate) of the output is a function with input from $\mathbf{R}$ and with output to $\mathbf{R}$. Thus r_1 is an "ordinary" function. So is r_2. A standard example of such a function is the position of an object on a flat plate as a function of time. Create some specific examples; include the constant function.

Finally, write down the definitions of limits in these special cases in the notation above, as well as in the most general notation.

Section 10.3.1

10.34. A sample diagram is below:

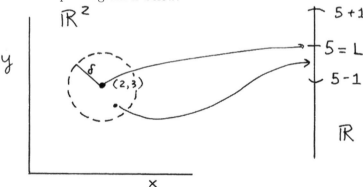

10.35. A suitable function with a limit is $r(t) = (t, t^2)$; for one without, put a standard (non-) example into the first coordinate, and make the second coordinate zero.

10.36. For the function g note that the resulting picture is almost exactly that for a one-variable function limit; if we restrict our attention to the y–z plane, it is as if we were graphing $z = f(y)$ and the picture is exactly as before. The addition of the variable x simply stretches things in the x direction, turning a horizontal strip into a horizontal slab. Note that the vertical strip is turned into *not* a vertical rectangular box, but a vertical cylinder.

If you have trouble with the last one, try graphing it so that the x–z

plane is in the plane of the paper, and the y axis is coming out at you. This reduces it to two dimensions, rather as we did with g just above. Now try the usual orientation.

Section 10.3.4

10.52. Clearly you can force $|x-1| < \delta$ and $|y-2| < \delta$ as before. To control $|xy - 1 \cdot 2|$ remember the trick we used before when coping with a product (then it was $f(x) \cdot g(x)$), namely adding and subtracting something. One can get $|xy-2| = |xy-y+y-2| \le |xy-y|+|y-2| = |y| \cdot |x-1| + |y-2| < |y| \cdot \delta + \delta$. There is one more standard trick, namely the "bounding" of y; if we were sure $\delta \le 1$, then $|y| < 3$, so the term we are really interested in is bounded above by $3\delta + \delta = 4\delta$. Fine, so $\delta = \min(1, \epsilon/4)$ ought to work. Show that it does.

But this has none of the simplicity of the "coordinatewise" approach for $f \colon \mathbf{R} \to \mathbf{R}^2$, and (e.g.) $\sin(xy)$ looks extremely difficult. Now what?

Section 10.3.5

10.57. The diagram might look something like this:

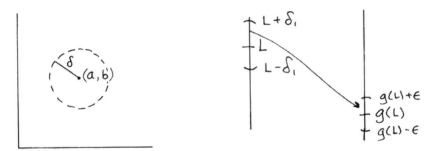

As for the proof, imitate work in Section 6.5.3. But simply take the "ϵ" required for $g \circ f$, insert it into the limit machine for g, yielding a δ_1. (More formally, insert ϵ into the definition of $\lim_{x \to L} g(x) = g(L)$, where you know that the limit is $g(L)$ since g is continuous at L.) Treat this δ_1 as an "ϵ" for f and produce a δ. This δ works for $g \circ f$.

Section 10.4.2

10.64. First, set the notation. Let $P = \{a = x_0, x_1, x_2, \ldots, x_{n-1}, x_n = b\}$, and let the maximum value of f in the interval $[x_{i-1}, x_i]$ be M_i. Then

$$U = M_1 \cdot (x_1 - x_0) + M_2 \cdot (x_2 - x_1) + \ldots + M_n \cdot (x_n - x_{n-1}).$$

Suppose the additional point in the partition is x' introduced between x_{j-1} and x_j. Then for the new sum we must find two new maxima, say, M_1' in $[x_{j-1}, x']$ and M_2' between $[x', x_j]$. The change in the sum from U to U' is the insertion of $z = M_1' \cdot (x' - x_{j-1}) + M_2' \cdot (x_j - x')$ to replace $w = M_j \cdot (x_j - x_{j-1})$.

How does M_j compare to M_1' and M_2'? Then get an inequality for z vs. w. Deduce the inequality for the sums.

Section 10.4.3

10.67. The number you would like to have around to be the l.u.b. is, of course, $\sqrt{2}$. Unfortunately, $\sqrt{2}$ is not in $\mathbf{Q}$ (the set of rational numbers), so is not present in the universe. An l.u.b. will be hard to find: think of the sequence $1, 1.4, 1.41, 1.414, \ldots$ of rational approximations to $\sqrt{2}$. If 1.41 is proposed as the l.u.b., it is easy to see that $1.414^2 < 2$, and so 1.41 is not even an upper bound. And 1.42 is not a least upper bound, for although it is an upper bound, so is 1.415.

These arguments are intuitive, but here is a real one (we step into $\mathbf{R}$ when convenient, and then retreat again into $\mathbf{Q}$). First, no rational number less than $\sqrt{2}$ can be an upper bound (so not the least upper bound), for if r is such a number, there is some rational number in the real number interval $(r, \sqrt{2})$, say s, and $s^2 < \sqrt{2}^2 = 2$, so s is in the set but $s > r$. Second, although any rational number greater than $\sqrt{2}$ is an upper bound, it is not a least upper bound; suppose $r \in \mathbf{Q}$, greater than $\sqrt{2}$, is an upper bound for our set. There is some rational number in the real interval $(\sqrt{2}, r)$, say s, and s is an upper bound for our set although it is less than r, contradicting the assumption that r is the l.u.b. .

There are sets in the universe of rational numbers that have a least upper bound ($\{r : r \in \mathbf{Q}$ and $r < 1\}$), but there are also sets that, although bounded above, do not have a least upper bound. This distinction between $\mathbf{Q}$ and $\mathbf{R}$ is what makes $\mathbf{R}$ the right set to do analysis on, actually.

10.68. The intuition might be that if u is the least upper bound of S, then there are points of S arbitrarily close to u, but less, as indicated by the picture:

Section 10.4.5

10.75. The statement about L vs. U for a particular partition is true, but has nothing to do with what we want. To say that each number x in a set S_1 has some number y in another set S_2 so that $x < y$ is nice, but is by no means to say that there is a single number greater than all numbers in S_1. For example, for odd integer n there is some even integer m so that $n < m$, but the collection of odd integers is not bounded above (and none of this "∞" stuff either: ∞ is not a number).

10.77. Suppose we have some partition P of $[0, 1]$, and form $U(f, P)$. In each of the subintervals of the partition, the maximum value of f is 1 (since there is always a rational number in the subinterval), and so $U(f, P) = 1 \cdot (x_1 - x_0) + 1 \cdot (x_2 - x_1) + \ldots + 1 \cdot (x_n - x_{n-1})$ (the sum of the lengths of the subintervals, i.e., the whole length of the interval, i.e., 1). (Alternatively, note that the sum above "telescopes.") So for any P, $U(f, P) = 1$. So the set $\mathcal{U}$ (usually containing lots of values) is actually what? So its g.l.b. is what? What about $\mathcal{L}$?

10.81. The picture might look as follows for some generic function f:

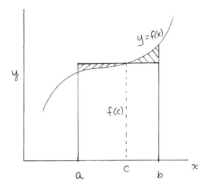

Note that one can actually see the perfect one rectangle partition. As for the proof, note that $\dfrac{\int_a^b f}{b - a}$ (a number) lies in the interval $[m, M]$ and apply the Intermediate Value Theorem for functions.

10.82. Surely the function has a maximum and minimum value on any subinterval, the minimum value always 0, the maximum value 1 if .5 is in the subinterval, and 0 if not. Lower sums always have value zero, and upper sums have value the length of the subinterval in which .5 lies.

To show f integrable we must show that some l.u.b. is equal to some g.l.b.[1] Since all the lower sums are zero, the l.u.b. of the lower sums is also zero, and $\int f = 0$ if f is indeed integrable. To show the g.l.b. of the upper sums is zero, find some partitions with associated upper sums as close to zero as you like.

Section 10.4.6

10.83. To show the fact about g.l.b.'s, try a proof by contradiction. Suppose that the g.l.b. for f is strictly smaller than that for g. Draw a number line picture; what must there be somewhere in the interval between them, by the g.l.b. version of Proposition 10.4.2? Because it is there, something else must occur, by the first thing you proved in the exercise. There's the contradiction.

10.86. Since f is continuous at c, there is some $\delta > 0$ so that for all x, if $|x - c| < \delta$ then $|f(x) - f(c)| < \epsilon$. Now let x be arbitrary such that $|x - c| < \delta$. In the case $x > c$ we know there is some u_x in $[c, x]$ so that

$$\left| \frac{F(x) - F(c)}{x - c} - f(c) \right| = |f(u_x) - f(c)|.$$

---

[1] Note that we don't, at the moment, have a complete characterization of which functions are Riemann integrable, although we know that some kinds are.

But since u_x is between c and x, it is closer to c than x, and so $|u_x - c| < \delta$. But then $|f(u_x) - f(c)| < \epsilon$, and so

$$\left| \frac{F(x) - F(c)}{x - c} - f(c) \right| = |f(u_x) - f(c)| < \epsilon,$$

as required. The case for $x < c$ is similar; consider the interval $[x, c]$.

The proof may be done without the Mean Value Theorem for Integrals, simply by using Proposition 10.4.7 and a slightly different argument.

References

[1] E. T. Bell. *Men of Mathematics*. Simon and Schuster, New York, 1937.

[2] C. H. Edwards, Jr. *The Historical Development of the Calculus*. Springer-Verlag, New York, 1979.

[3] Robert M. Exner. How can X possibly Equal 2? *New York State Mathematics Teachers Journal*, 11(2):42–46, 1961.

[4] Richard R. Goldberg. *Methods of Real Analysis (Second Edition)*. John Wiley & Sons, Inc., New York, 1976.

[5] George Pólya. *How to Solve It*. Princeton University Press, Princeton, NJ, 1945.

[6] Wolfram Research, Inc. *Mathematica© v. 2.2.2 for the Macintosh*. Wolfram Research, Champaign, Illinois, 1993.

Index

Undergraduate Texts in Mathematics

(continued from page ii)

Isaac: The Pleasures of Probability. *Readings in Mathematics.*

James: Topological and Uniform Spaces.

Jänich: Linear Algebra.

Jänich: Topology.

Kemeny/Snell: Finite Markov Chains.

Kinsey: Topology of Surfaces.

Klambauer: Aspects of Calculus.

Lang: A First Course in Calculus. Fifth edition.

Lang: Calculus of Several Variables. Third edition.

Lang: Introduction to Linear Algebra. Second edition.

Lang: Linear Algebra. Third edition.

Lang: Undergraduate Algebra. Second edition.

Lang: Undergraduate Analysis.

Lax/Burstein/Lax: Calculus with Applications and Computing. Volume 1.

LeCuyer: College Mathematics with APL.

Lidl/Pilz: Applied Abstract Algebra. Second edition.

Logan: Applied Partial Differential Equations.

Macki-Strauss: Introduction to Optimal Control Theory.

Malitz: Introduction to Mathematical Logic.

Marsden/Weinstein: Calculus I, II, III. Second edition.

Martin: The Foundations of Geometry and the Non-Euclidean Plane.

Martin: Geometric Constructions.

Martin: Transformation Geometry: An Introduction to Symmetry.

Millman/Parker: Geometry: A Metric Approach with Models. Second edition.

Moschovakis: Notes on Set Theory.

Owen: A First Course in the Mathematical Foundations of Thermodynamics.

Palka: An Introduction to Complex Function Theory.

Pedrick: A First Course in Analysis.

Peressini/Sullivan/Uhl: The Mathematics of Nonlinear Programming.

Prenowitz/Jantosciak: Join Geometries.

Priestley: Calculus: A Liberal Art. Second edition.

Protter/Morrey: A First Course in Real Analysis. Second edition.

Protter/Morrey: Intermediate Calculus. Second edition.

Roman: An Introduction to Coding and Information Theory.

Ross: Elementary Analysis: The Theory of Calculus.

Samuel: Projective Geometry. *Readings in Mathematics.*

Scharlau/Opolka: From Fermat to Minkowski.

Schiff: The Laplace Transform: Theory and Applications.

Sethuraman: Rings, Fields, and Vector Spaces: An Approach to Geometric Constructability.

Sigler: Algebra.

Silverman/Tate: Rational Points on Elliptic Curves.

Simmonds: A Brief on Tensor Analysis. Second edition.

Singer: Geometry: Plane and Fancy.

Singer/Thorpe: Lecture Notes on Elementary Topology and Geometry.

Smith: Linear Algebra. Third edition.

Smith: Primer of Modern Analysis. Second edition.

Stanton/White: Constructive Combinatorics.

Stillwell: Elements of Algebra: Geometry, Numbers, Equations.

Stillwell: Mathematics and Its History.

Stillwell: Numbers and Geometry. *Readings in Mathematics.*

Strayer: Linear Programming and Its Applications.

Thorpe: Elementary Topics in Differential Geometry.

Toth: Glimpses of Algebra and Geometry. *Readings in Mathematics.*